PROBIT ANALYSIS

PROBIT ANALYSIS

A STATISTICAL TREATMENT OF THE SIGMOID RESPONSE CURVE

BY

D. J. FINNEY, M.A.

Lecturer in the Design and Analysis of Scientific Experiment,
University of Oxford; Formerly of Rothamsted
Experimental Station

WITH A FOREWORD BY

F. TATTERSFIELD, D.Sc., F.R.I.C.

Head of the Department of Insecticides and Fungicides,
Rothamsted Experimental Station

CAMBRIDGE

AT THE UNIVERSITY PRESS

1947

CAMBRIDGE UNIVERSITY PRESS
Cambridge, New York, Melbourne, Madrid, Cape Town, Singapore, São Paulo, Delhi

Cambridge University Press
The Edinburgh Building, Cambridge CB2 8RU, UK

Published in the United States of America by Cambridge University Press, New York

www.cambridge.org
Information on this title: www.cambridge.org/9780521135900

First published 1947
This digitally printed version 2009

A catalogue record for this publication is available from the British Library

ISBN 978-0-521-13590-0 paperback

CONTENTS

LIST OF EXAMPLES

FOREWORD

IN our work at Rothamsted on insecticides, their action and relative potency, we have been so dependent and have made so many calls on the skill and patience of the Statistical Department and in particular, in recent years, of Mr Finney, that it is a pleasure to learn that the statistical methods and techniques which he has placed so willingly at our disposal for the solution of our various problems are now to be expressed in a more permanent form and to be given to a wider field of workers. If this book receives its due, the investigators of toxicological problems throughout the world will find it a standby.

To many of us, engaged on the practical issues of devising the experimental material and methods to be employed in our laboratory tests, there will be much in these pages that has to be taken upon trust. They are meant to act primarily as an aid in computation, but there is a profound thread of reasoning running through them all, giving a coherence to the several chapters. One cannot but feel, therefore, that the more mathematical readers will find the book suggestive and stimulating.

Twenty-five or more years ago, when I entered the field of research from which Mr Finney takes so many examples for detailed computational study, the very whisper of the need for statistical analysis, falling upon the ears of the biological expert, was enough to bring down a storm of denial upon one's head. Although there may be some small residue of such a reaction still in existence, it now only persists in obscure nooks and crannies of the world of biological research. Much of this change is due to the school of statisticians founded by R. A. Fisher at Rothamsted, of which Mr Finney has been a distinguished member. They always showed a willingness to enter into one's experimental difficulties and an understanding of the limitations imposed by time and space upon the amount of work it was possible to do. The measure of the thought given to these matters can be gauged by the fact that some of the most important tables, given by Mr Finney in the following pages, were in process of computation in actual anticipation of problems which we later carried to him. This brings up the now rather hackneyed, but none-the-less important, matter of the prior consultation with a competent statistician before designing experiments. Repeatedly is such a course justified by events, in that the plan can not only be often simplified, but so designed as to yield more information with little, if any, more labour.

Another issue, raised by such text-books as this, concerns the extent to which a training in statistics should enter into the curriculum of the biological- and biochemical student. I have found, relatively late in life, how hard a task it is to pick up, during busy years, the requisite amount of basic statistical knowledge to follow the arguments set out in recent text-books written for one's own benefit. I therefore feel that more attention might be given to such a matter by the academic powers-that-be. The engineer may not need to know all there is to be scientifically known about the composition and manufacture of his tools and materials, but at least he should know enough to use them rightly. So too the quantitative biological investigator.

Mr Finney's text-book has a very long and dignified ancestry ranging back to early Egyptian and Mesopotamian times, when texts of mathematics were used for the training of scribes, the professional computers of those distant days. But this is a forward-looking book and is primarily meant for experimentalists. Enlarged editions may well follow this, the first, as the subject grows and new problems arise; and that there are many just round the corner anyone who has discussed these matters with the author is very well aware. Having had the benefit of his personal advice and help throughout the last six years I cannot but wish this book Godspeed on its helpful mission to others.

F. TATTERSFIELD

PREFACE

FROM the theory of probability, originally investigated in order to explain nothing more important than the results of games of chance, has developed the science of applied statistics. Over one hundred years ago Laplace wrote that

... la théorie des probabilités n'est au fond, que le bon sens réduit au calcul: elle fait apprécier avec exactitude, ce que les esprits justes sentent par une sorte d'instinct, sans qu'ils puissent souvent s'en rendre compte,

and these words might equally well be written of statistics to-day. In many fields of scientific research, and especially in the biological sciences, numerical studies are complicated by the inherent variability of the material under investigation, and conclusions must be based on averages derived from series of observations. The estimation of these averages and the assessment of their reliability are statistical operations, in the performance of which the experimenter inevitably employs a statistical technique even though he himself may not always recognize this fact. The operations may be simple or complex, depending upon the circumstances; if, however, they cease to be 'le bon sens réduit au calcul', they can no longer be expected to contribute to the understanding of the problem under investigation.

The recent rapid advances in the application of rigorous statistical methods to biological data began with the publication, in 1925, of R. A. Fisher's *Statistical Methods for Research Workers*. Not only did Fisher develop exact methods for the analysis of data from small samples to replace the older approximations from large-sample theory, but he also introduced new and powerful techniques for making the most efficient use of experimental results. Of equal importance to the growth of the present-day philosophy of experimentation was Fisher's suggestion that the statistician should be consulted during the planning of an experiment and not only when statistical analysis of the results is required, as his advice on experimental design may greatly increase the value of the results eventually obtained. Since Fisher's book first appeared, the principles of experimental design and the methods of statistical analysis have been extended so rapidly as to make it increasingly difficult for any but the professional statistician to be familiar with the variety of methods needed in biological problems.

Many books since Fisher's have been written with the aim of surveying a wide field of biological statistics, but these can give only an outline of some important topics. There is to-day a need for books in which the specialized statistical methods appropriate to certain branches of science will be discussed in sufficient detail to enable biologists to appreciate them and apply them to their own problems.

One subject requiring fuller discussion than can reasonably be expected in any general text-book of statistics is the method of probit analysis, for the development of which J. H. Gaddum and C. I. Bliss are largely responsible. This method is widely used for the analysis of data from toxicity tests for the assay of insecticides and fungicides, and also of data from other types of assay dependent upon a quantal response. In this book I have tried to give a systematic account of the theory and practice of probit analysis, including as much as possible of the most recent extensions and refinements, in such a form that it may be understood by biologists, chemists, and others who have some knowledge of elementary statistical procedure; at the same time, I have endeavoured to satisfy the mathematical statistician by showing the theoretical background of the method. The less mathematically minded reader will no doubt be content to omit, or at most to read cursorily, Appendix II and other sections concerned with the mathematical basis of the technique. Full understanding and appreciation of statistical methods can be gained only by experience in their use, but careful study of the numerical examples should enable many who were previously unfamiliar with probit analysis to apply it satisfactorily to their own data.

This book has been written as a result of several years of close collaboration with members of the Insecticides Department at Rothamsted Experimental Station, especially with Dr F. Tattersfield, Dr C. Potter, and, until he left Rothamsted, Dr J. T. Martin. I wish to express my gratitude to them for discussing with me a wide variety of their problems, for advising me on the experimental aspects of their results, and for the generosity with which they have permitted me to use their data both in this book and in earlier publications. I am also very grateful to my colleagues in the Statistical Department at Rothamsted for much helpful discussion, and particularly to Dr F. Yates for his detailed and constructive criticism in the preparation of my book. Others to whom my thanks are due include Miss G. M. Ellinger for assistance in German translation, Dr C. G. Butler for permission to use the numerical data of Ex. 33, Dr A. E. Dimond and Dr J. G. Horsfall for giving me very full information on the results discussed in § 41 and for permission to use their data, Professor G. H. Thomson for assistance in tracing the history of the probit method, the Editors of the *Annals of Applied Biology* for permission to reproduce the first half of Table II, Professor R. A. Fisher, Dr F. Yates, and Messrs Oliver and Boyd, Ltd. for permission to reproduce Tables I, VI and VII from their book *Statistical Tables for Biological, Agricultural and Medical Research*, and my father, Robt. G. S. Finney, for very considerable help in the correction of proofs.

D. J. FINNEY

ROTHAMSTED EXPERIMENTAL
STATION

August 1945

ERRATA

p. 43.

[†] Fundamentable Table

should be

[†] Fundamental Table

Chapter 1

INTRODUCTORY

1. BIOLOGICAL ASSAY

THE term *biological assay*, in its widest sense, should be understood to mean the measurement of the potency of any stimulus, physical, chemical or biological, physiological or psychological, by means of the reactions which it produces in living matter. The biological method of measuring the stimulus is adopted either for lack of any alternative, or because an exact physical or chemical measurement of stimulus intensity may need translation into biological units before it can be put to practical use.

Biological assay is most commonly considered as referring to the assessment of the potency of vitamins, hormones, toxicants and drugs of all types by means of the responses produced when doses of these are given to suitable experimental animals. Estimation of the potency of a natural product, such as a drug extracted from plant material, in producing a biological effect of a certain type, is often impossible or impracticable by chemical analysis. Even if the chemical constitution of the material is known or determinable, there may be little knowledge of the magnitude of the effect which the constituents will produce, a difficulty not confined to natural products but occurring also with many manufactured compounds, such as insecticides, which are made to precise chemical specifications yet which are of unknown biological activity. The material must in fact be tested and standardized by methods appropriate to its future use.

For example, vitamin assays may be made in terms of weight changes or other physical measurements observed in rats, the effects of different doses of the preparation to be assayed being compared with the effects of a standard in order to estimate the relative potency of the test preparation and the standard. Insulin may be assayed in terms of the fall in blood sugar in injected rabbits, and digitalis by the mortality amongst injected cats. Again, the potency of insecticides may be assessed by means of the mortality in batches of treated insects, and that of fungicides by the proportion of treated spores failing to germinate. Another form of assay procedure which is sometimes useful depends on measurement of the time required for the production of a specified effect instead of measurement of the magnitude of the effect produced. In an interesting and informative article, which should be read by all who are seriously concerned with this type of investigation, Bliss and Cattell (1943) have reviewed nearly 300 recently published papers on the theory and practice of biological assay, with especial reference to vitamin, hormone, and drug assay. The texts of Burn (1937) and Coward (1938) may also be consulted, though the statistical methods there advocated do not fully exploit modern developments.

One type of assay which has been found valuable in many different fields, but especially in toxicological studies, is that dependent upon the *quantal*, or all-or-nothing, response. Though quantitative measurement of a response is always to be preferred when available, there are certain responses which permit of no graduation and which can only be expressed as 'occurring' or 'not-occurring'. The most obvious example of this kind of response is death; although workers with insects have often found difficulty in deciding precisely when an insect is dead (Tattersfield *et al.* 1925), in many investigations the only practical interest lies in whether or not a test insect is dead, or perhaps in whether or not it has reached a degree of inactivity such as is thought certain to be followed by early death. In fungicidal investigations, failure of a spore to germinate is a quantal response of similar importance. In studies of drug potency, the response may be the cure of some particular morbid condition, no possibility of partial cure being under consideration. This book is chiefly concerned with assays made by means of quantal responses, though in Chapter 10 some attention is given to quantitative responses. Most of the discussions are presented in terms of tests of the potency of insecticides and fungicides, since it is for these that the methods of analysis were first developed systematically; the same methods, however, are applicable to many other data, both biological and non-biological.

2. Variability of Responses

One feature possessed by all biological assays is the variability in the reaction of the test subjects and the consequent impossibility of reproducing at will the same result in successive trials, however carefully the experimental conditions are controlled. Though similar variability may be encountered in assays based only on purely physical or chemical measurements, it is generally then of far less practical importance. The contrast between the physical approach and the biological may be seen from a consideration of two methods for the estimation of the ratio of two unknown weights. The physical method is to balance each in turn against a set of standardized weights, and to take as the required estimate the ratio of their magnitudes. There may be technical difficulties in carrying out the operations of weighing to very high accuracy, and both the quality of the balance and the competence of the operator are important factors, but for most practical purposes the reproducibility of the results is not called in question; one measurement on each weight will usually suffice to determine the ratio with an accuracy far beyond that obtainable in any biological assay.

The physical assay of the ratio is here so simple that no alternative method is needed. For the sake of the illustration it may be compared with a biological technique, using quantal responses, in which the weights are dropped from a fixed height on to the heads of live rats. Data for the assay are provided by the records of death or survival. That the first weight, at its first trial, killed a rat, while the second weight did not, would not show with any certainty that the first was the heavier, still less would it give any clue to their ratio; the effect would be influenced not only by the weight dropped, but also by the age, sex, size and physical condition of the rat, and other biological and environmental factors (as well as, of course, the shape and elasticity of the weights, which will here be assumed the same for both). If batches of rats, chosen at random from the stock available, were tested with each weight, the proportionate effect of variation in susceptibility from rat to rat would be reduced with increasing size of sample, and the weights could be compared in terms of the two mortality rates. Variability could be still further controlled, though never entirely eliminated, by using a specially bred strain of rats, and selecting batches homogeneous for sex, age and other relevant factors. When every test is made from the same arbitrary height, this assay cannot discriminate between weights too light to cause any deaths or between weights so heavy as to kill every rat. This difficulty can be overcome by making tests from a series of different heights and obtaining a range of mortalities for each weight. The weights are then compared in terms of equivalent heights, or heights estimated to give the same (say 50 %) mortality. The height scale thus provides a basis for the biological comparison of any number of weights, but, without experimental or theoretical knowledge of the law relating mortality to height and the physical measure of weight, the results of the biological assay cannot be transformed to purely physical terms.

This example has been discussed in some detail, as, in spite of its absurdity, it illustrates the necessity for a careful consideration of variability in any biological assay technique. To some extent the quantal nature of the responses is a complication, but quantitative responses by no means provide an escape from the problem. Equal doses of insulin will not produce equal effects on the blood sugar of different rabbits, or even on the blood sugar of the same rabbit at different times. Consequently, though two insulin preparations could be compared in terms of the magnitudes of the changes in blood sugar produced in two rabbits, only repetition of the tests on several rabbits for each preparation can give an estimate of the relative potency sufficiently precise to be of any practical value.

Biological aspects of, and reasons for, variability in test organisms of many kinds have been discussed by Clark (1933, especially Chapter VI), and his remarks on individual variations in response deserve careful reading. The occurrence of this variability introduces considerations other than those of biology; when there is a large natural variability of response amongst the test subjects, the analysis of numerical data for the estimation of the effects of applied treatments can only be effected satisfactorily with the aid of exact statistical techniques.

3. STATISTICAL METHODS

The development of statistical techniques for the analysis of biological data of all types has proceeded with great rapidity in recent years. In many fields of research on biological topics, experimental and observational results can only be used to the best advantage by subjecting them to precise and critical statistical examination. When a programme of biological research involves the collection of numerical data, the problem of interpreting these is almost inevitably one of statistics. The choice is not, as the biologist sometimes imagines, of whether his figures shall be 'statistically analysed' or not, but rather of whether the analysis shall be theoretically sound and able to extract all the relevant information from the material, or inadequate and possibly unsound. Even the simplest and most straightforward averaging of results is essentially a statistical process; the analysis appropriate to any body of data is determined by the inherent properties of those data, not by the whim of the statistician. It is unfortunate, to say the least, that good experimental work should ever be followed by a statistical treatment of the results so unsatisfactory that the conclusions are incomplete, unreliable, or even actively misleading.

The function of the statistician in biological investigations is to supply that critical and objective judgement of numerical material which is a product of his specialized training and experience. An important aspect of his work is co-operation in the planning of an experimental programme so that, taking into account all relevant information already available, it is designed to give results of maximum utility and precision. The assistance of a competent statistician from the beginning of the programme will often substantially increase the value of the results obtained from a given amount of experimental time and labour, in respect of both their scope and their reliability, whereas the conclusions may be much less satisfactory if the statistician is only consulted after the completion of the experimental work.

Nevertheless, the methods of analysis used by the statistician are not esoteric mysteries, but are simply instruments for discovering the most important features of numerical data. The computational procedures appropriate to many types of data have been so far standardized that they can be applied by a biologist who has some understanding of their purposes, even though he may know little of their theoretical foundations. The blind application of formulae is a danger which should be avoided, for not infrequently the formulae may be used quite inappropriately; on the other hand, the anxiety of many biologists to learn enough of statistical methods in order to be able to analyse their own data without complete dependence on the assistance of a statistician is witnessed by the recent spate of books designed to instruct the non-mathematician in statistical technique.

The statistical treatment of quantal assay data has been much aided by the development of *probit analysis*. This method, which is usually attributed to Gaddum (1933) and Bliss (1934a, b; 1935a, b) though it has, in fact, a much longer history (§ 14), has now been widely adopted as the standard method of reducing the data to simple terms.

4. SUMMARY OF CONTENTS

This book is written with the intention of introducing the probit method to many who have previously not ventured to use it, and of presenting some of its more recent developments to those who are already familiar with it. In the first few chapters the technique is shown in its simplest form, stripped of all but the essentials. It is hoped that these chapters, at least, will be capable of appreciation and use by many whose knowledge of other branches of statistics is small. Even for this purpose, however, a slight acquaintance with modern statistical thought and terminology is necessary, and although notes on various tests and distributions will be found in the appropriate sections these can do little more than give references and hints on particular applications. The reader is strongly recommended to familiarize himself with the relevant portions of R. A. Fisher's *Statistical Methods for Research Workers* (1944), especially the sections dealing with the normal, t, and χ^2 distributions, and with regression. K. Mather's *Statistical*

Analysis in Biology (1943) provides a valuable introduction for those who find Fisher's book too difficult.

The numerical examples in subsequent chapters have been carefully chosen to illustrate many points of procedure and to show the application of the method to a variety of toxicological data. Though the computational work required is sometimes laborious, it is not as heavy as some accounts have made it appear; in an appendix is given a detailed description of a systematic arrangement of the computations for the simplest type of problem, and this arrangement may easily be extended to suit more complex data. A second appendix gives a brief outline of the mathematical theory of the probit method. The book is completed by a series of tables which lessen considerably the computing time and labour required for probit analysis.

Chapter 2

QUANTAL RESPONSES AND THE DOSAGE-RESPONSE CURVE

5. THE FREQUENCY DISTRIBUTION OF TOLERANCE

IN all biological assays there are two components to be considered, the *stimulus* (for example, a vitamin, a drug, a physical force, or a mental test) and the *subject* (for example, an animal, a plant, a piece of tissue, or a single cell). The stimulus is applied to the subject at an intensity specified in units of concentration, weight, time, or other appropriate measure and under environmental conditions as carefully controlled as is practicable, as a result of which a *response* is produced by the subject. Different stimuli are then compared in terms of the magnitudes of the responses they produce, or, more commonly and usefully, in terms of the intensities required to produce equal responses.

When the characteristic response is quantal, its occurrence or non-occurrence will depend upon the intensity of the stimulus applied. For any one subject, under controlled conditions, there will be a certain level of intensity below which the response does not occur and above which the response occurs; in psychology such a value is designated the *threshold* or *limen*, but in pharmacology and toxicology the term *tolerance* seems more appropriate. This tolerance value will vary from one member to another of the population used, frequently between quite wide limits. When the characteristic response is quantitative, the stimulus intensity needed to produce a response of any given magnitude will show similar variation between individuals. In either case, the value for an individual also is likely to vary from one occasion to another as a result of uncontrolled internal or external conditions. Clark (1933, Chapter VI) has discussed the nature of these individual variations in response for many different populations.

For quantal response data it is therefore necessary to consider the distribution of tolerances over the population studied. If the *dose*, or intensity of the stimulus, is measured by λ, the distribution of tolerances may be expressed by

$$dP = f(\lambda)\,d\lambda; \qquad (2\cdot1)$$

this equation states that a proportion, dP, of the whole population consists of individuals whose tolerances lie between λ and $\lambda + d\lambda$, where $d\lambda$ represents a small interval on the dose scale, and that dP is the length of this interval multiplied by the appropriate value of the *distribution function*,* $f(\lambda)$.

If a dose λ_0 is given to the whole population, all individuals will respond whose tolerances are less than λ_0, and the proportion of these is P, where

$$P = \int_0^{\lambda_0} f(\lambda)\,d\lambda; \qquad (2\cdot2)$$

the measure of dose is here assumed to be a quantity which can conceivably range from zero to $+\infty$, response being certain for very high doses so that

$$\int_0^{\infty} f(\lambda)\,d\lambda = 1.$$

The distribution of tolerances, as measured on the natural scale, may be markedly skew, but it is often possible, by a simple transformation of the scale of measurement, to obtain a distribution which is approximately *normal*. 'A variate is said to be normally distributed when it takes all values from $-\infty$ to $+\infty$ with frequencies given by a definite mathematical law, namely, that the logarithm of the frequency at any distance d from the centre of the distribution is less than the logarithm of the frequency at the centre by a quantity proportional to d^2. The distribution is therefore symmetrical, with the greatest frequency at the centre; although the variation is unlimited, the frequency falls off to exceedingly small values at any considerable distance from the centre, since a large negative logarithm corresponds to a very small number' (Fisher, 1944, § 12). In tests of insecticidal sprays, for example, although the distribution of tolerance concentration of the toxic agent is usually far from symmetrical on

---

* The statement that $f(\lambda)$ is a function of λ means simply that for any given value of λ the value of $f(\lambda)$ is uniquely determined.

account of a few insects with extremely high tolerances providing an extended 'tail' to the distribution (Fig. 1), normalization can often be effected by expressing the tolerances in terms of the logarithms of the concentrations instead of the absolute values (Fig. 2); this transformation is now accepted as standard practice for expressing the results of such trials (cf. Galton, 1879). Various writers (Clark, 1933; Hemmingsen, 1933; Bliss, 1935a) have sought an explanation of the normal distribution of log tolerances in the Weber-Fechner law and in adsorption phenomena, particularly as expressed by the Langmuir adsorption law, but these explanations are beyond the scope of this book. The validity and appropriateness of the logarithmic transformation in the analysis of experimental data are not dependent on the truth or falsity of any hypotheses relating to adsorption; use of the log concentration as measuring the dosage in insecticidal trials requires no more justification than that it introduces a simplification into the analysis. There are additional advantages in having a scale on which a given proportionate increase in concentration has the same scale value at all levels of concentration, but other forms of transformation may sometimes be more suitable. Parker-Rhodes (1941, 1942a, b) has advanced reasons for expecting a normal distribution of some fractional power of the concentration of a fungicide to which suspensions of fungus spores are exposed (see § 45), though this must be only an approximation which holds over a restricted range of concentrations.

It is convenient to take x as representing the intensity of the stimulus on the scale on which the tolerances are normally distributed, and λ as the untransformed value of concentration, time of exposure, or other variate. Thus for much insecticidal work, if λ is the concentration of the toxic agent,

$$x = \log_{10}\lambda, \tag{2.3}$$

and for some fungicides a better transformation may be

$$x = \lambda^i, \tag{2.4}$$

where usually $i \leqslant 1$. The second normalizing transformation tends to the logarithmic as i is decreased to zero. There is no reason why a simple transformation should always be available; nevertheless, many classes of data have been found amenable to treatment on

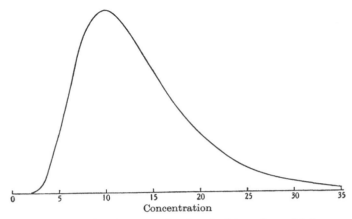

FIG. 1. Typical distribution curve for the absolute values of tolerance concentrations of an insecticide. (The area between any two ordinates represents the proportion of insects having tolerances lying between these limits.)

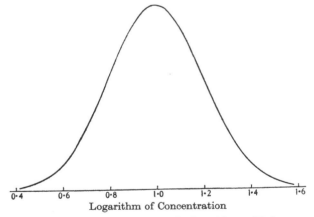

FIG. 2. Normal distribution for the logarithms of tolerance concentrations, derived from Fig. 1.

these lines, so that the study of the consequences of this normal distribution and of the appropriate methods of statistical analysis is of considerable practical importance. In order to distinguish the scales of measurement, the word *dose* will be restricted to the scale of λ, in which measurements of stimulus are in fact made, and x will be referred to as the measure of dosage, or, more briefly, just the *dosage*.

In an investigation for which tolerance can be satisfactorily defined, so that for any given dose all individuals with equal or lower tolerance values will respond, a graph of the percentage responding against the dose will give a steadily rising curve. The rate of increase in response per unit increase in dose is frequently very low in the region of zero or 100 % response, but higher in the intermediate region, so that the curve is sigmoidal (Fig. 3). When the stimulus is measured in dosage units, the curve takes the characteristic *normal sigmoid* form (Fig. 4). This curve does not attain the zero or 100 % response except at infinitely low or infinitely high dosage, a situation which cannot truly arise (except that, when the measure of dosage intensity is logarithmic, an infinitely low value represents zero dose). Nevertheless, the distribution may be effectively normal over the range of values which is of practical interest, the disagreement between theory and fact outside this range being of negligible importance.

6.   DIRECT MEASUREMENT OF TOLERANCE

The tolerance of the test subject in respect of a given stimulus can sometimes be measured directly. Such direct measurement, for example, is involved in the 'cat' method for the assay of digitalis, in which anaesthetized cats are given a continuous slow intravenous infusion of digitalis until death occurs. If there is any appreciable time lag between the introduction of the drug and its taking effect, the lethal dose will be overestimated. Though there is no certainty that the dose required to cause death under conditions of slow application will be the same as the tolerance for more rapid application, the technique has proved suitable for assaying a preparation of unknown potency in terms of a standard.

An alternative method is to give to each subject successive doses of different intensities, allowing a suitable time interval

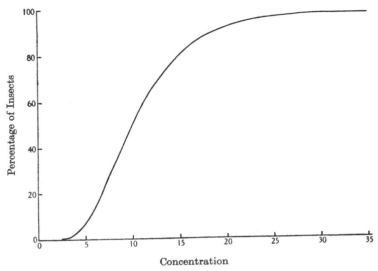

FIG. 3. Sigmoid curve derived from Fig. 1 to show percentage of insects whose tolerances are less than a specified value.

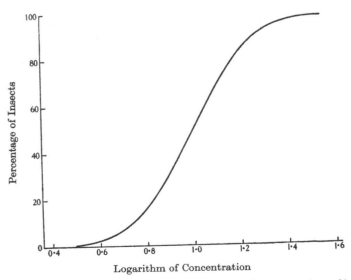

FIG. 4. Normal sigmoid curve derived from Fig. 2 to show percentage of insects whose log tolerances are less than a specified value.

after each for a return to normal and making the differences sufficiently small for a satisfactory determination of the lowest dose which causes the characteristic response. With an irreversible response, such as death, the doses would have to be given in an increasing series. For the method to be satisfactory, there must be no cumulative effect of doses already given, either as lowering or increasing the resistance of the subject, a condition which severely limits its applicability.

In either of these methods of direct measurement of tolerance, the appropriate methods of statistical analysis are the same as for other types of biological measurements. If the tolerance of each subject has been separately and independently determined, the set of values obtained may be subjected to the same analytical processes as measurements of length or weight; the estimation of means and standard errors, the comparison of distributions, and the making of tests of significance present no new features. Bliss and Hanson (1939), for example, have discussed the application of the analysis of variance and covariance to assays based on the 'cat' method.

Direct tolerance measurement is often impracticable on account of the amount of time required for either of the methods mentioned above. Even more frequently it is ruled out entirely by the nature of the problem. It is hard to conceive of any direct measurement technique for the poison tolerance of an insect, still less for that of a fungus spore. In these circumstances an entirely different approach must be adopted, and the potency of the stimulus must be assessed by means of the proportion of subjects, in random samples of the population, showing responses at different levels of dose.

7. THE BINOMIAL DISTRIBUTION

If an insect, selected at random from a population, is exposed to a dose λ_0 of a poison, the probability that it will respond is P; the probability of its failing to respond is $(1 - P)$, a quantity usually denoted by Q. The dose here may be measured by the concentration of toxic substance, the absolute quantity used, the length of time of exposure to a fixed set of conditions, or some combination of these or other factors. If two insects are exposed to the dose, and if their reactions are completely independent, the probability that both respond is P^2, and the probability that both fail to respond is Q^2; the probability that only the first responds is $P \times Q$, and the probability that only the second responds is $Q \times P$. Thus the total probabilities of 2, 1 and 0 responding are P^2, $2PQ$ and Q^2 respectively, the successive terms in the expansion of $(P+Q)^2$. In a similar manner it may be seen that if a batch of n insects is exposed to the dose λ_0, and all react independently, the probabilities of n, $(n-1)$, $(n-2)$, ..., 2, 1, 0 responding are the $(n+1)$ terms in the expansion of the binomial $(P+Q)^n$. The probability of exactly r responding is therefore

$$\frac{n!}{r!(n-r)!} P^r Q^{n-r}.$$

This is known as the *Binomial Distribution* of probabilities (cf. Fisher, 1944, § 18; Mather, 1943, § 5). The average number responding in repeated batches of n from the same population is nP, and the average number failing to respond nQ. Durham *et al.* (1929) have given useful tables of sums of terms from this distribution, and Clopper and Pearson (1934) have shown similar results in the form of charts.

The reactions of separate members of a batch to the stimulus of a particular dose are not always independent; a correlation of response may result from incomplete randomness of selection of the batch, or alternatively from unsatisfactory control of experimental conditions causing the number responding to be seriously affected by some factor other than the dose. For example, if each batch consists of insects from a single brood, insects from one batch are likely to be more alike in tolerance than those from different batches, and the variation between batches in the numbers responding will be greater than that for the binomial distribution. Again, the susceptibility of insects to an insecticide might be greatly influenced by temperature; if the temperature during the tests was permitted to vary substantially from one batch to another, the variance of the numbers responding to a given dose of insecticide would be inflated. The extreme situation is that, in every batch tested, either all members respond or all fail to respond, so that the evidence from a batch is no more reliable than that from an individual. Whatever the cause, such heterogeneity must make the weight to be attached to the data less than is appropriate to the binomial distribution (Bliss, 1935a; Parker-Rhodes, 1941).

The result of testing a series of doses, each on a separate batch of insects, is to obtain for each dose a proportion, p, of insects in the batch which show the characteristic response and whose tolerances are therefore lower than that dose. Each value of p is an estimate of the corresponding P, the proportion in the population of which the batch was a sample, and it is from these quantities that the statistics* of the population may be calculated. In general both P and p will increase steadily with increasing dose (an interesting exception is discussed in § 41), but, if the number of test subjects in a batch is small, sampling variation may interfere with the regularity of the trend in p. Trevan (1927) has shown that if two batches of five subjects are given doses which would cause 25 % and 75 % of responses respectively in the whole population, only 92 % of trials would give more responses for the higher dose. In 2 % of trials the lower dose would appear to be the more effective, in nearly 6 % the numbers of responses in the two groups would be equal, and in a very small proportion, 0·05 %, either none or all would respond to both doses. The larger the batches the greater is the assurance that there will be satisfactory discrimination between the effects of different doses, but when, as is often the case, the limiting factor to the size of the experiment is the total number of subjects to be used, it is usually preferable to have several batches of moderate size than to have two or three large ones, in order that a wide range of doses may be tested and an idea of the dose-response relationship obtained.

---

* The word *statistic* is here used in the sense introduced by Fisher (1944, § 11) as 'a value calculated from an observed sample with a view to characterising the population from which it is drawn'.

8. The Median Effective Dose

At one time it was customary to characterize the effectiveness of a stimulus by means of the *minimal effective dose*, or, for a more restricted class of stimuli the *minimal lethal dose*, terms which fail to take account of the variation in tolerance within a population. Writing of toxicity tests, Trevan (1927) says: 'The common use of this expression in the literature of the subject would logically involve the assumptions that there is a dose, for any given poison, which is only just sufficient to kill all or most of the animals of a given species, and that doses very little smaller would not kill any animals of that species. Any worker, however, accustomed to estimations of toxicity, knows that these assumptions do not represent the truth.' It might be thought that the minimal lethal dose of a poison could instead be defined as the dose just sufficient to kill a member of the species with the least possible tolerance, and also a *maximal non-lethal dose* as the dose which will just fail to kill the most resistant member. Though there will undoubtedly be doses so low that no test subject will succumb to them and doses so high as to prove fatal to all, there are considerable difficulties in the way of determining the end-points of these ranges. Even when the tolerance of individuals can be measured directly, to say, from measurements on a sample of ten or a hundred, that the lowest tolerance found indicated the minimal lethal dose would be unwise; a larger sample might well contain a more extreme member. When only quantal responses for selected doses can be recorded the difficulty is increased, and the occurrence of exceptional individuals in the batches at different dose levels may seriously bias the final estimates. The problem is, in fact, that of determining the dose at which the sigmoid death curve for the whole population meets the zero or 100 % levels of kill and a very extensive experiment would be necessary in order to estimate these points with any accuracy.

As a characteristic of the stimulus which can be more easily determined and interpreted, Trevan has advocated the *median lethal dose*, or as a more general term to include responses other than death, the *median effective dose*. This is defined as the dose which will produce a response in half the population, and thus, from another point of view, is the mean tolerance. If direct measurement of tolerance were possible the mean tolerance of a batch of test subjects would naturally be considered as the chief characteristic of the dose, and there is a strong case for using an estimate of the same quantity in material of the type now under discussion. The median effective dose may conveniently be referred to as the ED 50, and the more restricted concept of median lethal dose as the LD 50. Analogous symbols may be used for doses effective for other proportions of the population, ED 90, for example, being the dose which causes 90 % to respond. As will become apparent in later chapters, by experiment with a fixed number of test subjects, effective doses in the neighbourhood of ED 50 can usually be estimated more precisely than those for more extreme percentage levels, and this characteristic is therefore particularly favoured in expressing the effectiveness of the stimulus; its chief disadvantage is that in practice, especially in toxicological work, there is much greater interest attaching to doses producing nearly 100 % responses than to those producing only 50 %, in spite of the difficulty of estimating the former.

For any distribution of tolerances, the ED 50, Λ, satisfies the equation

$$\int_0^\Lambda f(\lambda)\,d\lambda = 0.5. \qquad (2.5)$$

When a simple normalizing transformation for the doses is available, so that x, the normalizing measure of dosage, has a normally distributed tolerance, equation (2·1) is transformable to

$$dP = \frac{1}{\sigma\sqrt{(2\pi)}}\,e^{-\frac{1}{2\sigma^2}(x-\mu)^2}dx, \qquad (2.6)$$

where μ is the centre of the distribution and σ^2 its variance. Thus μ is the population value of the mean dosage tolerance, or median effective dosage, and efforts must be directed at estimating it from the observational data. This problem will be considered at length in Chapter 3. For the present the normalizing transformation will be assumed logarithmic, as defined by equation (2·3), so that μ is the log ED 50; the results obtained are in the main true for any other transformation, at least as far as they relate to the measure of dosage, x, but modifications are required in transforming back from the x to the λ scale.

Clearly the ED 50 alone does not fully describe the effectiveness of the poison or other stimulus tested. Two poisons may require the same rates of application in order to be lethal to half the population, but, if the distribution of tolerances has a lesser 'spread' for one than for the other, any increase or decrease from this rate will produce a greater change in mortality for the first than for the second. This 'spread' is measured by the variance, σ^2: the smaller the value of σ^2, the greater is the effect on mortality of any change in dose. Stimuli which produce their effects by similar means (in particular, poisons whose physiological effects are similar), often have approximately equal variances of their log tolerances for any given population of test subjects, even though they differ substantially in their median lethal doses. An assessment of the relative potencies can then be made from median lethal doses alone (§ 20).

Chapter 3

THE ESTIMATION OF THE MEDIAN EFFECTIVE DOSE

9. THE PROBIT TRANSFORMATION

A TYPICAL test used in the evaluation of an insecticide is one in which successive ˙batches of insects are exposed to different concentrations of the poison for a constant time and, after a suitable interval, scored for the numbers dead and alive. As an alternative to varying the concentration, a fixed concentration may be used throughout, but different total quantities given. Another factor which is sometimes studied at different levels is the period of exposure, the concentration and quantity of poison being kept constant. Such experimental conditions have the character of the stimuli discussed in § 5 whose effects are observed at different levels of intensity. Statistical methods for the analysis of quantal response data have been developed in recent years chiefly for use with tests of this type.

The form of analysis now used to estimate the parameters μ and σ^2 of the distribution of tolerances, equation (2·6), is generally based upon the *probit transformation* of the experimental results. The history of this transformation and of the statistical technique associated with it is outlined in § 14; Bliss (1934 b) first proposed the name 'probit' for his modification of Gaddum's normal equivalent deviate, which he increased by 5 so as to simplify the arithmetical procedure by avoiding negative values. The probit of the proportion P is defined as the abscissa which corresponds to a probability P in a normal distribution with mean 5 and variance 1; in symbols, the probit of P is Y, where

$$P = \frac{1}{\sqrt{(2\pi)}} \int_{-\infty}^{Y-5} e^{-\frac{1}{2}u^2} du. \qquad (3\cdot1)$$

The effect of transformation from percentages or proportions to probits is illustrated in Fig. 5. The normal sigmoid curve of

Fig. 4 is reproduced here, together with the straight line obtained when its ordinates are replotted on a linear scale of probits. Along the left-hand vertical axis is a linear scale of percentages with their corresponding probit values, and on the right-hand

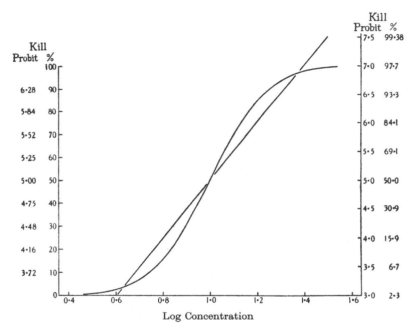

FIG. 5. Effect of the probit transformation. (The normal sigmoid curve of Fig. 4 is transformed to a straight line when the ordinates are measured on a linear scale of probits instead of percentages.)

axis is a linear scale of probits with their corresponding percentage values. The transformation may be considered as a stretching of the left-hand scale to give that on the right-hand, during which process the sigmoid curve becomes straightened.

If (2·6) represents the distribution of tolerances on the x scale of dosages, the expected proportion of insects killed by a dosage x_0 is

$$P = \frac{1}{\sigma \sqrt{(2\pi)}} \int_{-\infty}^{x_0} e^{-\frac{1}{2\sigma^2}(x-\mu)^2} dx.$$

Comparison of the two formulae for P then shows that the probit of the expected proportion killed is related to the dosage by the linear equation

$$Y = 5 + \frac{1}{\sigma}(x - \mu). \tag{3.2}$$

By means of the probit transformation, experimental results may be used to give an estimate of this equation, and the parameters of the tolerance distribution may then be estimated; in particular, the median effective dosage is estimated as that value of x which gives $Y = 5$.

TABLE 1.   Transformation of Percentages to Probits

%	0	1	2	3	4	5	6	7	8	9
0	—	2·67	2·95	3·12	3·25	3·36	3·45	3·52	3·59	3·66
10	3·72	3·77	3·82	3·87	3·92	3·96	4·01	4·05	4·08	4·12
20	4·16	4·19	4·23	4·26	4·29	4·33	4·36	4·39	4·42	4·45
30	4·48	4·50	4·53	4·56	4·59	4·61	4·64	4·67	4·69	4·72
40	4·75	4·77	4·80	4·82	4·85	4·87	4·90	4·92	4·95	4·97
50	5·00	5·03	5·05	5·08	5·10	5·13	5·15	5·18	5·20	5·23
60	5·25	5·28	5·31	5·33	5·36	5·39	5·41	5·44	5·47	5·50
70	5·52	5·55	5·58	5·61	5·64	5·67	5·71	5·74	5·77	5·81
80	5·84	5·88	5·92	5·95	5·99	6·04	6·08	6·13	6·18	6·23
90	6·28	6·34	6·41	6·48	6·55	6·64	6·75	6·88	7·05	7·33
	0·0	0·1	0·2	0·3	0·4	0·5	0·6	0·7	0·8	0·9
99	7·33	7·37	7·41	7·46	7·51	7·58	7·65	7·75	7·88	8·09

A table giving probits for specified values of P has been prepared by Bliss (1935a), and this table is reproduced by Fisher and Yates (1943) as Table IX of their *Statistical Tables for Biological, Agricultural and Medical Research*. A simplified version of this table, sufficiently detailed for many purposes, is given as Table 1 above, and the full table is reproduced as Table I. The relationship between percentages and probits is shown graphically in Fig. 6.

10.   THE PROBIT REGRESSION LINE

When experimental data on the relationship between dose and mortality have been obtained, either a graphical or an arithmetical process can be used to estimate the parameters. Both depend upon the probit transformation. The graphical approach is much more rapid and is sufficiently good for many purposes, but for some more complex problems, or when an accurate assessment of the precision of estimates is wanted, the more detailed arithmetical analysis is necessary. In this chapter only

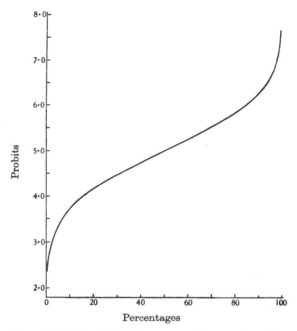

FIG. 6. Relationship between percentages and probits.

the graphical method will be discussed, though the ideas introduced will be wanted again for the discussion of the maximum likelihood estimation of parameters in Chapter 4.

In order to make either type of estimate, the percentage kill observed for each dose must first be calculated and converted to probits by means of Table 1.* The probits are then plotted against x, the logarithm of the dose (or against some other

---

* There is seldom any advantage in using Table I for these empirical probits; even when batches of test subjects are sufficiently large to justify the use of three or more places of decimals (in itself a rare occurrence), points cannot usually be plotted with this accuracy.

normalizing function of the dose where this seems more suitable), and a straight line drawn by eye to fit the points as satisfactorily as possible. In drawing the line and judging its agreement with the data, only the vertical deviations of the points must be considered: the line must be so placed that the differences between the probit values which are plotted and the probits given by the line at the same dosages are as small as possible. Very extreme probits, say outside the range 2·5–7·5, carry little weight and may almost be disregarded unless many more insects were used than in the batches giving intermediate probit values. The line is, in statistical terminology, the weighted *regression line* of the mortality probit on x.

This line may be used, as described in § 17, to initiate the arithmetical process of estimating a better fitting line. The empirical probits plotted for a carefully conducted experiment often lie so close to a straight line, however, that there is no necessity to improve on the provisional line. Only experience of the subject and of the experimental technique used can be a sound guide in this matter, but it is undoubtedly true that many experimenters who make use of probit analysis spend time unnecessarily on arithmetic when eye estimation would suffice. As will appear from § 17, the complete method for deriving the best estimate of the line is not difficult, but is laborious if adopted as a routine measure for all tests made.

If it is decided to proceed with the eye estimate alone, the log LD 50 is estimated from the line as m, the dosage at which $Y = 5$. The slope of the line, b, which is an estimate of $1/\sigma$, is obtained as the increase in Y for a unit increase in x. These two parameters are then substituted in equation (3·2) to give the estimated relationship between dosage and kill. To test whether the line is an adequate representation of the data, a χ^2 test (Fisher, 1944, § 20) may be used, as in Ex. 1 below. A value of χ^2 within the limits of random variation indicates satisfactory agreement between theory (the line) and observation (the data). A significantly large χ^2 may arise either because the individual test subjects do not react independently to the poison, or because the straight line does not adequately describe the relation between dosage and probit. In the former case the scatter of the points about the line will be wider than would occur if there were no correlation between the reactions of insects in the same batch; the precision of the line will be reduced (§ 11), though its position should be free from bias providing that adequate precautions have been taken in the conduct of the experiment (§ 42). In the latter case there will generally be a systematic departure of the points from the line, indicating a curvilinear relationship; it may be possible to transform this to a linear relationship by adopting a different scale of dosage, such as that of equation (2·4).

A timely warning against attaching too much importance to the probit itself, at the expense of the kill, has been given by Wadley (Campbell and Moulton, 1943), who says: 'The use of transformations carries with it a temptation to regard the transformed function as the real object of study. The original units should be mentioned in any final statement of results.' In essence the probit is no more than a convenient mathematical device for solving the otherwise intractable equations discussed in Appendix II. Though it may also be used to give a simple diagrammatic representation of the dosage-response relationship, and though familiarity enables these diagrams to be interpreted directly, any suggestion that the statistical analysis is completed by the estimation of a probit regression line must be avoided.

Many of the numerical examples in this book have been chosen to illustrate special points of analytical technique, and, since the data have been removed from their original context, their discussion may not always be carried as far as a statement of conclusions in biological or chemical terms. In practical applications of the probit method the results should finally be expressed by median effective doses, relative potencies, tolerance variances, or other suitable quantities, the units employed being dose (not log dose or dosage) and percentage kill; at this stage the word 'probit' need seldom be mentioned, though sometimes quantities such as mean probit differences (§ 24) may usefully be retained.

Ex. 1.   *Fitting by eye of a probit regression line to the results of an insecticidal test.* Martin (1942, Table 9) has published data showing the effect of a series of concentrations of rotenone when sprayed on *Macrosiphoniella sanborni*, the chrysanthemum aphis, in batches of about fifty. His results are reproduced in Table 2. The number affected, shown in the third column of the table, is the total of insects apparently dead, moribund, or so badly affected as to be unable to walk more than a few steps. This classification has been found convenient and has been frequently used by Tattersfield and his co-workers at Rothamsted (Tattersfield *et al.*, 1925); the total is taken as the 'kill', and normal or only slightly affected insects are considered to have survived.

TABLE 2. Toxicity of Rotenone to *Macrosiphoniella sanborni*

Concentration (mg./l.)	No. of insects (n)	No. affected (r)	% kill (p)	Log concentration (x)	Empirical probit
10·2	50	44	88	1·01	6·18
7·7	49	42	86	0·89	6·08
5·1	46	24	52	0·71	5·05
3·8	48	16	33	0·58	4·56
2·6	50	6	12	0·41	3·82
0	49	0	0	—	—

The rotenone was applied in a medium of 0·5 % saponin, containing 5 % of alcohol. Insects were examined and classified one day after spraying.

In experiments of this nature, provision should always be made for estimating the natural mortality amongst untreated insects. The last line of Table 2 records that forty-nine insects were sprayed with the alcohol-saponin medium alone, containing no rotenone, and that all ˙of these survived. It therefore seems safe to assume that the results for the five concentrations of rotenone have not been appreciably influenced by the superimposition of a natural mortality of insects. As will be seen later (Chapter 6), adjustments to the statistical analysis are needed when there are indications of an appreciable natural mortality during the course of an experiment.

The percentage kills observed at each concentration are shown in the fourth column of Table 2. These values are estimates of the corresponding averages, P, for the whole population, and are subject to sampling errors. In order to distinguish them from the population values they will be denoted by p. (In all formulae in the text, here and elsewhere, p and P denote proportions, not percentages; it is convenient to use the same symbols in the headings of tables both for proportions and percentages, though for the latter $100p$ and $100P$ would be more correct. If this is borne in mind, no confusion should be caused by the slight ambiguity of usage.)

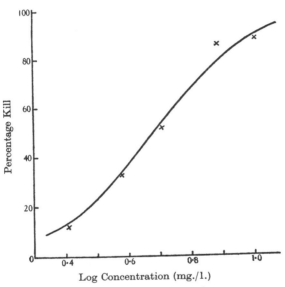

FIG. 7. Relationship between dosage of rotenone and percentage kill of *M. sanborni* (Ex. 1), showing normal sigmoid curve.

Over the range of concentrations tested, the sigmoid nature of the relationship between percentage kill and log concentration is not very apparent. The percentages are plotted against the dosage in Fig. 7, together with the normal sigmoid curve which is fitted to them by the present analysis. Between 25 % and 75 % kill this curve is practically indistinguishable from a straight line; a line drawn to fit the five points would give a dosage of about 0·68 corresponding to 50 % kill, a value in good agreement with that obtained later for the log LD 50. The straight-line relationship for percentages must nevertheless be quite inadequate at more extreme values, even though the absence of extremes and the regularity of the data enable it to be used here. A method proposed by Kärber (see § 13 below) for the estimation of the LD 50 is dependent upon this linear approximation.

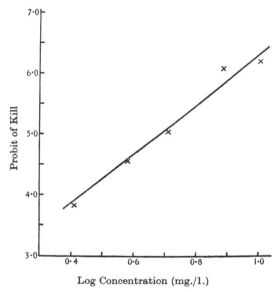

FIG. 8. Relationship between dosage of rotenone and probit of kill of *M. sanborni* (Exs. 1 and 6), showing probit regression line.

The probits of the percentage kills, read from Table 1, have been entered in the last column of Table 2. These probits are plotted against dosage in Fig. 8; they lie nearly on a straight line, and such a line has been drawn through them by eye. Using Table 1 once more, points on the straight line of Fig. 8 have been converted back from probits to percentages; these give the sigmoid curve already shown in Fig. 7, from which the mortality to be expected from any given dosage, or the dosage which will kill, on the average, a given percentage of insects, may be predicted. In practice the sigmoid seldom needs to be constructed, as all predictions can be made directly from the probit diagram.

For example, in Fig. 8 a probit value of 5 is given by a dosage of $m = 0.687$; this therefore is the estimate of log LD 50, and the LD 50 is estimated as a concentration of 4·86 mg./l. The value may be compared with 4·85 mg./l. obtained by the arithmetical process of fitting a straight line to the same data (Ex. 6). Similarly the log LD 90 corresponds to a probit of 6·28 and is therefore 1·003; the LD 90 is thus estimated as 10·1 mg./l.

Again from Fig. 8, an increase of 0·8 in dosage is associated with an increase of 3·21 in the probit.[*] Hence the estimated

TABLE 3.   Comparison of Observed and Expected Mortality in Eye Estimation for Rotenone-*Macrosiphoniella sanborni* Test

Log concentration (x)	Y	P	No. of insects (n)	No. affected Observed (r)	No. affected Expected (nP)	Discrepancy $(r-nP)$	$\dfrac{(r-nP)^2}{nP(1-P)}$
1·01	6·30	90·3	50	44	45·2	−1·2	0·33
0·89	5·83	79·7	49	42	39·1	2·9	1·06
0·71	5·10	54·0	46	24	24·8	−0·8	0·06
0·58	4·58	33·7	48	16	16·2	−0·2	0·00
0·41	3·90	13·6	50	6	6·8	−0·8	0·11
							$\chi^2_{[3]} = 1.56$

regression coefficient of probit on dosage, or the rate of increase of probit value per unit increase in x, is

$$b = 1/s = 4.01,$$

where $s\,(= 0.25)$ is an estimate of σ, the standard deviation of the distribution of log tolerances. The relationship between probit and dosage may be written

$$Y = 5 + 4.01(x - 0.687), \quad \text{or} \quad Y = 2.25 + 4.01x. \quad (3.3)$$

Equation (3·3) may be used to calculate expected numbers of insects killed at each concentration. By substitution of the values

[*] De Beer (1941) has suggested the graduation of a protractor in units of b instead of degrees, so as to read the slope of the line directly; the same instrument could be used for all diagrams having a fixed ratio between the scale units of probits and of dosage.

of x used in the experiment, the equation gives the values of Y shown in the second column of Table 3; from Table 1 the corresponding expected percentage kills, P, are obtained. Thus a probit of 6·30 corresponds to a percentage of between 90 and 91, or, more accurately, $90 + \frac{2}{6}$ %. If the expected proportion for any concentration is multiplied by n, the number of insects tested at that concentration, the result is the *expected number* of affected insects, or the average number which would be affected in a batch of size n if equation (3·3) represented the true relationship between dosage and kill. These numbers, nP, may then be compared with the actual numbers affected, r, in order to judge the adequacy of the equation.

Only the second concentration tested shows any appreciable discrepancy, three more insects being affected than equation (3·3) predicts, and even this may be shown to be well within the limits of random variation. There is no indication of any systematic departure from the line such as might arise if the wrong normalizing transformation had been used and the true equation were not linear. A test of significance of the discrepancies may be obtained by squaring each, dividing the square by $(1 - P)$, and again dividing by the tabulated value nP.* The sum of these quantities is, to a sufficiently close approximation if the line in Fig. 8 has been well drawn, a χ^2; the degrees of freedom are two less than the number of concentrations tested, since the two parameters of equation (3·3) have been estimated from the data. The mean value of χ^2 in random sampling from a population whose tolerance distribution is defined by this equation is equal to the number of degrees of freedom; the value obtained in Table 3 is 1·56, which, being less than 3, is clearly sufficiently small to be attributed to random fluctuations about the relationship specified in (3·3). Indeed Fisher and Yates's Table IV, of which a simplified form is reproduced as Table VI, shows that a value greater than 7·8 could occur by chance in 5 % of cases; hence the probit regression line in Fig. 8 appears to be a very satisfactory representation of the results of the experiment.

* This calculation, like that of nP, may be performed on a slide rule with sufficient accuracy for the present purpose.

11. PRECISION OF THE ESTIMATES

The binomial distribution of probabilities (§ 7) shows that if a batch of n insects is exposed to a concentration which, on an average, will kill a proportion P, the probabilities of n, $(n-1)$, ..., 2, 1 being killed are the successive terms in the expansion of $(P+Q)^n$, provided that the n insects react independently to the poison. By a well-known result of elementary statistical theory, it follows that the standard deviation of the number of affected insects in samples of size n is $\pm\sqrt{(nPQ)}$, and that the standard deviation of the observed proportion p, about its mean value of P, is $\pm\sqrt{(PQ/n)}$. The square of a standard deviation is known as the *variance*; here the variance of the proportion killed is PQ/n, which is inversely proportional to the number in the batch. The reciprocal of the variance, sometimes called the *invariance* or *quantity of information*, is proportional to n, and represents the weight to be attached to the observation on the batch in respect of the information it provides on P. In Appendix II the weight to be attached to the probit of P is shown to be nw, where

$$w = Z^2/PQ; \qquad (3\cdot4)$$

here Z is the ordinate to the normal distribution corresponding to the probability P, and may be written

$$Z = \frac{1}{\sqrt{(2\pi)}} e^{-\frac{1}{2}(Y-5)^2}. \qquad (3\cdot5)$$

Bliss (1935a) and Fisher and Yates (1943, Table XI) have tabulated the *weighting coefficient*, w, for values of Y at intervals of $0\cdot1$; intermediate values may be obtained by interpolation if required.* A shortened version of their table is given here as Table 4; more accurate values appear in the column for $C = 0$ of Table II. The values are symmetrical about $Y = 5\cdot0$, so that, for example, $Y = 3\cdot8$ and $Y = 10 - 3\cdot8 = 6\cdot2$ have the same weighting coefficient, $0\cdot370$. For many practical purposes Table 4

* A table of w at intervals of $0\cdot01$ in Y has been prepared by Bliss and reproduced by the Department of Pharmacology, Food and Drug Administration, U.S. Department of Agriculture, Washington, D.C.

is sufficiently accurate; it is very seldom necessary to interpolate for Y, which may usually be taken correct to the nearest 0·1.

The weighting coefficients may be used in order to estimate the standard error of the log ED 50. For this purpose the value of Y corresponding to each dosage used must be read from the provisional regression line as drawn by eye, the weighting coefficient determined from Table 4 for each Y, and this coefficient multiplied by n, the number of insects tested at the dosage. The quantities nw must then be summed for all dosages; the symbol

TABLE 4.   The Weighting Coefficient, $w = Z^2/PQ$

Y	0·0	0·1	0·2	0·3	0·4	0·5	0·6	0·7	0·8	0·9
1	0·001	0·001	0·001	0·002	0·002	0·003	0·005	0·006	0·008	0·011
2	0·015	0·019	0·025	0·031	0·040	0·050	0·062	0·076	0·092	0·110
3	0·131	0·154	0·180	0·208	0·238	0·269	0·302	0·336	0·370	0·405
4	0·439	0·471	0·503	0·532	0·558	0·581	0·601	0·616	0·627	0·634
5	0·637	0·634	0·627	0·616	0·601	0·581	0·558	0·532	0·503	0·471
6	0·439	0·405	0·370	0·336	0·302	0·269	0·238	0·208	0·180	0·154
7	0·131	0·110	0·092	0·076	0·062	0·050	0·040	0·031	0·025	0·019
8	0·015	0·011	0·008	0·006	0·005	0·003	0·002	0·002	0·001	0·001

S is used to denote the summation. If the log ED 50 is not very different from the mean value of the dosages used in the experiment, its standard error is approximately $\pm 1/b\sqrt{(Snw)}$; this expression makes no allowance for the sampling errors in the estimation of b, the slope of the line, in consequence of which it may be a serious underestimation if m, the estimated log ED 50, is far from $\bar{x}$, the weighted mean dosage or $(Snwx)/(Snw)$. The variance of b is, by equation (II, 11), $1/Snw(x-\bar{x})^2$, and a more accurate value for the variance of m is therefore

$$V(m) = \frac{1}{b^2}\left\{\frac{1}{Snw} + \frac{(m-\bar{x})^2}{Snw(x-\bar{x})^2}\right\}. \qquad (3\cdot6)$$

Equation (3·6) gives the variance of the logarithm of any ED value by substitution of its estimate for m. In making a rapid analysis of data, by means of the provisional line only, inclusion of the second term in the variance of m is often unnecessary; its calculation requires part of the complete calculation for fitting a more accurate regression line, and if the computations have to be taken so far not a great deal of extra labour is needed for the more exact result. Litchfield and Fertig (1941) have described a more rapid, but less exact, approximate method which may sometimes be useful.

This discussion of weights, standard errors and variances is based on the assumption that there is no heterogeneity of departures of the plotted points from the regression line. If the reactions of the individuals in a batch are not independent of one another, the weights, nw, though still proportional to the true weights, will be too large, and the estimated variances will therefore be too small. This will be indicated by a large value of the statistic χ^2 (§ 10), which is now seen to be a weighted sum of squares of the discrepancies between the expected and observed numbers killed. Since the expected value of χ^2 is its number of degrees of freedom, a significantly large value indicates that all weights have been overestimated by a factor $\chi^2/(k-2)$, where k is the number of dosages tested. All variances should therefore be multiplied by this *heterogeneity factor* as compensation for the overweighting.

When χ^2 is not greatly in excess of its expectation and it may be assumed that the weights are correct without further adjustment, the standard errors may be considered in relation to a normal distribution of errors and fiducial limits calculated with the aid of normal deviates (Ex. 2 and § 12). On the other hand, a significant χ^2 necessitates the estimation of standard errors empirically from the observed variability between batches, and these errors must therefore be used in conjunction with the t-distribution (Fisher, 1944, § 23; Mather, 1943, § 15). The fact that the standard errors themselves are not precisely known leads to a wider range of values being admissible as within the limits of experimental variation for m (Exs. 7 and 9). Fisher and Yates (1943, Table III) have tabulated the distribution of t, the deviate required for this situation, and a simplified version of their table is given as Table VII; t must be used with the same number of degrees of freedom as in the χ^2. The values rapidly approach those of the normal distribution as the number of degrees of freedom increases.

Ex. 2. *The standard error of a median lethal dose.* The foregoing remarks may be illustrated on the data ·of Ex. 1. In Table 5 the values of Y as predicted by equation (3·3) for each concentration of rotenone are tabulated to the nearest 0·1. From Table 4 the w corresponding to each Y is entered and multiplied

TABLE 5.   Calculation of Standard Error of log LD 50 in Eye Estimation for Rotenone-*Macrosiphoniella sanborni* Test

Log concentration (x)	No. of insects (n)	Y	w	nw	nwx
1·01	50	6·3	0·336	16·8	16·968
0·89	49	5·8	0·503	24·6	21·894
0·71	46	5·1	0·634	29·2	.20·732
0·58	48	4·6	0·601	28·8	16·704
0·41	50	3·9	0·405	20·2	8·282
				119·6	84·580

$$\bar{x} = 0·7072, \quad (Snwx)^2/Snw = 59·81418,$$
$$Snwx^2 = 64·42700, \quad Snw(x - \bar{x})^2 = 4·61282.$$

by the corresponding n to give the column nw; again a slide rule gives sufficient accuracy. The sum of this column is

$$Snw = 119·6,$$

whence the standard error of m is, approximately, since χ^2 was found to be well within the limits of random variation,

$$s_m = 1/4·01 \sqrt{(119·6)}$$
$$= 0·0228.$$

The mean dosage, $\bar{x}$, may be found by constructing a column in Table 5 for nwx, summing, and dividing by Snw. This gives

$$\bar{x} = 84·580/119·6$$
$$= 0·707,$$

a value which differs but slightly from $m = 0·687$. In order to calculate the full expression for the variance of m, the entries in the column nwx must be multiplied again by x and added to give
$$Snwx^2 = 64·427.$$

The sum of squares of deviations of x may then be calculated as

$$Snw(x - \bar{x})^2 = Snwx^2 - (Snwx)^2/Snw$$

$$= 64 \cdot 427 - 59 \cdot 814$$

$$= 4 \cdot 613,$$

and therefore, by equation (3·6),

$$V(m) = \frac{1}{b^2} \left\{ \frac{1}{119 \cdot 6} + \frac{(0 \cdot 687 - 0 \cdot 707)^2}{4 \cdot 613} \right\}$$

$$= \frac{0 \cdot 008448}{(4 \cdot 01)^2} = 0 \cdot 000525.$$

The square root of this is the revised value for the standard error of m

$$s_m = 0 \cdot 0229,$$

which differs by a negligible amount from the approximation.

Now m is measured on a logarithmic scale (logarithms to base 10). The standard error of the LD 50, the antilogarithm of m, may be shown to be given by the formula

$$\text{S.E.}(10^m) = 10^m \times \log_e 10 \times s_m, \qquad (3 \cdot 7)$$

so that $$\text{S.E.(LD 50)} = 4 \cdot 86 \times 2 \cdot 30 \times 0 \cdot 0229$$

$$= 0 \cdot 26.$$

Thus the LD 50 is estimated as $4 \cdot 86 \pm 0 \cdot 26$ mg./l. This standard error is of little use, as the distribution of errors may be far from normal on the concentration scale, and tests of significance or fiducial limits (§ 12) should always be evaluated on the dosage scale. For example, if it is desired to test whether the LD 50 estimated for this experiment differs significantly from a true value of $4 \cdot 2$ mg./l., the difference should be examined in logarithmic units. The logarithm of the latter dose is $0 \cdot 623$, so that the estimate is greater by $0 \cdot 064 \pm 0 \cdot 023$, a deviation which is $2 \cdot 8$ times its standard error. The last line of Table VII, which gives deviates for the normal distribution, shows the probability of so large a deviation to be less than 1 % (more exactly, from Table I, the probability is about $0 \cdot 5$ %); hence the experiment has given an estimate of the LD 50 significantly greater than 4·2 mg./l.

If the standard error of estimation of the log LD 90 is required, the approximate formula breaks down, since the estimate is very different from $\bar{x}$, and equation (3·6) must be used. The variance is

$$V(\log \mathrm{LD}\,90) = \frac{1}{b^2}\left\{\frac{1}{119\cdot6} + \frac{(1\cdot003 - 0\cdot707)^2}{4\cdot613}\right\}$$

$$= 0\cdot001701,$$

so that the log LD 90 is estimated as $1\cdot003 \pm 0\cdot041$.

12. FIDUCIAL PROBABILITY

When a parameter such as the median lethal dose has been estimated from experimental data it is natural to wish to infer within what limits its true value may reasonably be expected to lie. A statement about the probability of the true value lying between certain limits cannot be made in terms of the ordinary concept of probability, by the use of which probabilities can be assigned only to statements about the occurrence of observations or of statistics calculated from the observations. In order to overcome this difficulty, the concept of *fiducial probability* has been introduced (Fisher, 1942, §§ 62, 63; 1944, § 2; Mather, 1943, § 58).

There is said to be a fiducial probability, F, that the true value of a parameter is greater than a specified value if that value is the lowest value from which the observation is not significantly different at the F level of probability. Similarly, there is said to be a fiducial probability, F, that the true value lies between specified upper and lower limits if the lower limit is the lowest value and the upper limit the highest value which would not be contradicted by a significance test at the $\frac{1}{2}F$ probability level; these are termed *fiducial limits* to the value of the parameters. Their meaning and calculation will be made clear by considering fiducial limits in the estimation of the LD 50 for the data of Ex. 1.

Ex. 3. *Fiducial limits to the LD50 in the rotenone*-Macrosiphoniella sanborni *test.* In Ex. 2 the log LD 50 for the data on the toxicity of rotenone to *M. sanborni* was estimated as

$$m = 0 \cdot 687 \pm 0 \cdot 023.$$

No evidence of significant heterogeneity of the points about the regression line was found, and consequently standard errors may be considered in relation to the normal distribution. From the last line of Table VII, there is a 5 % probability that the deviation from the true log LD 50 shall be at least 1·96 times the standard error, or a $2\frac{1}{2}$ % probability that the estimate shall be at least $1 \cdot 96 s_m$ ($= 0 \cdot 045$) less than the true value and a $2\frac{1}{2}$ % probability that the estimate shall be at least 0·045 greater than the true value. In other words, if the true log LD 50 were 0·732, estimates as low as or lower than 0·687 would occur in only $2\frac{1}{2}$ % of trials such as that under consideration and if the true log LD 50 were 0·642, estimates as high as or higher than 0·687 would occur in only $2\frac{1}{2}$ % of trials. There is therefore a 95 % fiducial probability that the true value lies between 0·732 and 0·642, which limits correspond to concentrations of 5·40 and 4·39 mg./l. If the limits had been estimated directly from the LD 50 and its standard error as $4 \cdot 86 \pm 1 \cdot 96 \times 0 \cdot 26$ mg./l., or 5·37 and 4·35 mg./l. they would have been equidistant from the estimated LD 50, whereas the limits derived on the logarithmic scale are not symmetrically placed on the concentration scale. The difference is trivial here, but with relatively larger standard errors, or for more extreme limits, it may be considerably greater. The method just discussed is itself an approximation, sufficiently good for many data; the exact procedure for finding fiducial limits is set out later (§ 19).

13. APPROXIMATE METHODS OF ESTIMATION

Before passing to the consideration of the arithmetical technique of fitting probit regression lines, other approximate methods for estimating the LD 50 which have achieved some popularity may be briefly mentioned. All these have been developed independently of the probit technique and are based on an assumed linearity of the relationship between the percentage kill and the dosage instead of between the mortality probit and the dosage. They may often be adequate, but they are likely to be misleading if the curvature of the dosage-response curve is appreciable over the range of dosages tested, especially if in addition the distribution of the dosages tested is markedly unsymmetrical about the log LD 50.

(a) *Method of extreme lethal dosages.* This method is intended for use in the limited class of experiments in which subjects are tested singly, one at each of a series of dosages; the interval, d, between successive dosages must be constant, and enough levels must be tested to cover the range from those practically certain to be ineffective to those practically certain to kill. The log LD 50 is estimated as the mean of the highest dosage which fails to kill and the lowest which kills. When d is of about the same magnitude as σ, the standard deviation in the distribution of log tolerances, this estimate, $m_{(a)}$ has the approximate standard error (Gaddum, 1933)

$$s_{m_{(a)}} = 0 \cdot 75 \sqrt{(\sigma d)}. \tag{3.8}$$

This standard error is of little use, however, as the easiest way of estimating σ is from a probit diagram, and if this is employed the LD 50 may as well be estimated from it.[*]

(b) *Behrens's method* is intended for experiments in which the dosages are equally spaced, and in which equal numbers of test subjects are used at each of a series of dosages covering the whole range of kills from zero to 100 %. For each dosage two quantities are calculated:

$S_{x-}(r) =$ the total number of individuals killed at dosages less than or equal to x,

$S_{x+}(n-r) =$ the total number of individuals surviving at dosages greater than or equal to x.

The estimate of log LD 50 is $m_{(b)}$, the value of x for which

$$S_{x-}(r) = S_{x+}(n-r), \tag{3.9}$$

[*] Application of the probit method to experiments in which only one individual is tested at each concentration is discussed in § 43.

and is obtained by interpolation where necessary. The method was suggested by Behrens (1929), from whose data, and from additional results of his own, Gaddum (1933) found an approximation for the standard error of $m_{(b)}$ in the form

$$s_{m_{(b)}} = 0 \cdot 81 \sqrt{(\sigma d/n)}. \tag{3.10}$$

This expression also is of little use, since knowledge of the value of σ can only be derived from a more elaborate analysis.

(c) *Kärber's method* (Kärber, 1931) is more adaptable than the two preceding approximations, as it does not require a constant interval between successive dosages nor even a constant number of test subjects at each dosage, though the number should not vary very widely. Once again the dosages tested must cover the whole, or practically the whole, range from zero to 100 % kill. The method assumes the whole of the increase in proportion killed between dosages x_i and x_{i+1}, $(p_{i+1} - p_i)$, to be attributable to a dosage $\frac{1}{2}(x_i + x_{i+1})$, and estimates the log LD 50 as the mean dosage. If k dosages are tested, and if $p_1 = 0$, $p_k = 1$ (that is to say the two extreme dosages kill respectively none and all the test organisms), the estimate is

$$m_{(c)} = \frac{1}{2} S(p_{i+1} - p_i) (x_{i+1} + x_i), \tag{3.11}$$

the summation being taken over all dosages. When the interval between successive dosages has a constant value, d, this estimate may be written more simply as

$$m_{(c)} = x_k - \frac{1}{2} d S(p_i + p_{i+1}). \tag{3.12}$$

The standard error of the estimate in equation (3.12) is

$$s_{m_{(c)}} = d \sqrt{[S(PQ/n)]}, \tag{3.13}$$

where P, Q are the expected proportions of killed and survivors at each dosage, for which smoothed value of p, q may be taken (Irwin and Cheeseman, 1939a). When the extremes of mortality are not reached within the range of dosages tested, formula (3.12) may sometimes be made applicable by the assumption that the next lower or next higher dosage at the end of the range would kill none or all respectively. The standard error for the estimate

in equation (3·11) may be found similarly, but the more complicated formula will not be given here. Epstein and Churchman
(1944) have discussed the theory of Kärber's method, but failure
to distinguish between the observed and expected proportions
detracts greatly from the usefulness of their work.

Gaddum (1933) has found that an average value of (3·13) may
be obtained in the same form as (3·8) and (3·10); the average
variance of m may be shown to be

$$V(m_{(c)}) = \sigma d/n \sqrt{\pi},$$

and therefore $s_{m_{(c)}} = 0{\cdot}751 \sqrt{(\sigma d/n)}.$ (3·14)

The variances of these three estimates of the log LD 50 may
be compared with that of the estimate obtained by the probit
method. According to Gaddum (1933), when dosages are equally
spaced and cover the whole range from 0 to 100 % kill, average
values of the variance are:

<blockquote>

Method of extreme lethal dosages: $0{\cdot}57\sigma d,$

Behrens's method: $0{\cdot}66\sigma d/n,$

Kärber's method: $0{\cdot}564\sigma d/n,$

Probit method: $0{\cdot}554\sigma d/n.$

</blockquote>

These values can be used only if σ either is known or can be
assumed the same as in earlier experience of the experimental
technique. The indications they give of the relative efficiencies
of the four methods, and in particular their suggestion of only
a trivial advantage for the probit method over that of Kärber
(amounting to a 2 % reduction in the variance of the estimated
log LD 50), are entirely misleading; as Gaddum has pointed out,
these conclusions apply only when the dosages are chosen in
such a way as to permit the use of Kärber's method or of one of
the other approximations. If previous experience or preliminary
trials give any clue to the value of the LD 50, the procedure of
distributing test subjects evenly over a wide range of dosages,
in order to ensure having zero or complete kills at the extremes,
is very wasteful. By concentrating on dosages nearer to the
LD 50, observations of much greater weight are obtained and

a correspondingly more precise estimate of the LD 50 is derived. A further theoretical objection to the three approximate methods is that, unless the dosages used are symmetrically situated with respect to the true log LD 50, each gives a biased estimate. Irwin (1937) found the bias to be negligible, even when the interval between successive dosages was as great as 2σ; the bias increases with this interval, but intervals greater than 2σ are not likely to occur frequently in practice.

Kärber's method, and to a lesser extent other approximate methods distinct from probit analysis, may occasionally be of use in aiding the rapid estimation of the LD 50, particularly in preliminary or exploratory trials. But the argument usually advanced in their favour, namely that they are much quicker than probit analysis and do not require its complex calculations, is of much less force than might at first appear. The graphical estimation of log LD 50 described in § 10 is as quick as either Kärber's or Behrens's methods, and has the important advantages of being just as easy and satisfactory to use when the dosages are not equally spaced or do not cover the whole range and when the numbers of subjects tested are not equal for all dosages. If assessment of the precision of the estimate of LD 50 is also required, some form of probit analysis is almost essential in order to estimate σ, and equation (3·13) is certainly more troublesome to use than the method given in § 11. By contrast with the other approximate methods, the graphical probit method avoids considerations of whether conditions for the applicability of Kärber's or Behrens's formulae are fulfilled, permits estimation of other properties of the dose-response relationship (such as LD 90) as easily as that of LD 50, and may easily be converted into the more exact analysis discussed in Chapter 4 when this seems necessary. Furthermore, as will be seen in later chapters, the probit method can be applied to many more complex types of toxicity test data.

Examples of the use of Kärber's method have been given by Irwin and Cheeseman (1939a, b); Ex. 4 below discusses very briefly the application of approximate methods to the data used in Ex. 1.

Ex. 4.   *Estimation of the median lethal dose by Behrens's and Kärber's methods.* In the experiment whose results are summarized in Table 2 the intervals between successive dosages did not differ greatly from the mean of 0·15, and the number of insects tested at each dosage was about 50. If an assumption is made that the next higher dosage would have killed all the insects and the next lower would have killed none, both Behrens's and Kärber's methods can be used for estimating the log LD 50. For the former the percentages dead and alive should be used rather than the actual numbers, in order to overcome the complication of unequal values of n. The two estimates are

$$m_{(b)} = 0·682, \quad m_{(c)} = 0·688.$$

The close agreement with the estimate made graphically in Ex. 1 is largely fortuitous; it occurs because the middle one of the five doses tested is very near to the LD 50 and the other doses and percentage kills are practically symmetrical on either side of this. Thus conditions happened to be very favourable to the two approximations, but even so they have no advantage over the graphical method as they can scarcely be reached with greater ease or rapidity than the estimate made in Ex. 1.

14.   HISTORY OF THE PROBIT METHOD

Though the widespread use of the probit transformation in the statistical analysis of biological data is of comparatively recent growth, the underlying principle has been known for many years. It appears to have originated with psychophysical investigators who, in the latter half of the nineteenth century, were confronted with the problem of estimating the magnitude of a stimulus from statements by test subjects that it seemed to them greater than or less than the various members of a standard series. The proportion of answers 'greater than' steadily decreases as the scale of standard stimuli is ascended, and shows a sigmoid relationship with the measure of these stimuli.

Fechner (1860) discussed the relationship of the difference between two weights with the proportion of trials in which a subject correctly judged which was the heavier. He suggested (Chapter VIII) the conversion of the proportions to deviates of a normal distribution with mean zero and precision* unity, by means of a table of the normal integral.[†] If steps are taken to eliminate biases due to the order, in time or space, in which the weights are picked up, when the weight difference is negligibly small the proportion of right answers should be one-half, and the normal deviate therefore zero. Fechner further suggested that a linear relationship would be found between the weight difference and the normal deviate. Hence, if the proportion of right answers were known for one weight difference, the factor of proportionality with the normal deviate could be estimated; estimates could then be made of the proportions corresponding to any other weight differences or vice versa. This appears to be the earliest reference to the fundamental idea of the probit method, namely, the reduction of a sigmoid response curve to a straight line by means of a transformation of the responses based on the normal integral; the credit for inventing the method should therefore be given to Fechner.

Müller (1879) recognized that the transformation from proportions to the standardized normal deviates introduced a differential weighting.[‡] He proposed to determine the parameters of the distribution of threshold values by fitting a straight line to the transformed data, weighting each point by its 'Müller weight', a quantity proportional to z^2 in the present notation. The ordinate, z, was taken as corresponding to the observed proportion, and not, as in equation (3·4), to the expected value from the fitted line. His method was called the *Constant Process* or the *Method of Right and Wrong Cases*, under which names it is still known to psychophysicists.

* The *precision*, $h = 1/\sigma\sqrt{2}$, was at one time commonly used instead of σ as a parameter of the normal distribution.

[†] This table is still known to psychophysicists as 'Fechner's Fundamentable Table'.

[‡] There is a danger of some verbal confusion here, since weights are used as stimuli, and, in the statistical analysis, the responses are assigned 'weights' proportional to their invariances.

Urban (1909, 1910) collected extensive experimental data, of the type considered by Fechner, on the difference threshold for lifted weights. In his experiments, a standard weight of 100 g. was lifted before each of a series of seven weights ranging from 84 to 108 g., and a judgement of lighter than, equal to, or heavier than, was given by the subject for each trial. Urban described several different methods for estimating the threshold of just perceptible weight differences, and for assessing the relationship between the weight difference and the proportion of right answers. In his account of the constant process, or, as he termed it, the $\Phi(\gamma)$ *Process*, he pointed out that the Müller weights failed to take account of the variability in the original proportions due to the binomial distribution of right and wrong answers. To allow for this source of variation, he introduced a factor $1/pq$, and weighted his normal deviates proportionally as z^2/pq.

Thomson (1914, 1919) drew attention to certain defects in Urban's treatment of the problem, especially in his methods of estimating the standard errors of the parameters and of assessing the goodness of fit of the line to the data. His revision (Brown and Thomson, 1940, Chapter III) put the constant process into a form very similar to that of the probit method to-day, except that the weights were taken from the values of p for the observations and not from the corresponding values of P for a provisional line, and that his formulae for the standard errors of the parameters were much more complicated and laborious to compute than those now used. No provision was made for taking account of zero or 100 % values of P, which were presumably simply ignored.

Independently of the work of the psychophysicists, Hazen (1914) and Whipple (1916) suggested the use of graph paper the scale of ordinates of which is graduated according to a normal probability distribution, so that the proportions are plotted as their corresponding normal deviates. A normal sigmoid curve plotted on this paper is automatically transformed to a straight line. A modified form of paper has a logarithmic scale of abscissae, so that a logarithmic transformation of dose is also made automatically at the same time.* O'Kane *et al.* (1930) made use of probability paper for plotting the results of insecticide tests.

In 1923 Shackell suggested that the normal integral might be used in interpreting the results of toxicity tests. Three years later Wright (1926), also without knowledge of the earlier work, proposed the use of an inverse function of the normal probability integral as a means of simplifying the statistical treatment of certain statistical data. His paper, however, seemed to escape the attention of biologists, who therefore had to wait a further seven years for yet another rediscovery of the method.

In 1933 Gaddum published an important memorandum on the analysis of quantal assay data in biological investigations. He proposed to transform each percentage to its *normal equivalent deviation* (N.E.D.), defined as the abscissa to a normal curve of zero mean and unit variance corresponding to a probability P; that is to say, P is the probability of obtaining an observation from this normal distribution whose value is less than or equal to the N.E.D. In symbols

$$P = \frac{1}{\sqrt{(2\pi)}} \int_{-\infty}^{\text{N.E.D.}} e^{-\frac{1}{2}u^2} du,$$

a transformation essentially the same as that used by Fechner. The N.E.D. of the percentage kill of various animals was found by Gaddum to give a straight line when plotted against the log dose of the drug applied. He described the regression technique for fitting the line, in a form similar to that given earlier by Urban and Thomson, but his treatment of the standard errors of the parameters and associated quantities was much simpler.

Bliss (1934*a*) suggested the division of the interval between 0·01 and 99·99 % into units of normal deviation which he called *probits*, the whole interval ranging from 0 to 10 probits and 50 % being 5 probits. When he later saw Gaddum's publication, he modified his definition of the probit and redefined it as in equation (3·1), so that it became the N.E.D. increased by 5 (Bliss, 1934*b*). In two comprehensive papers (1935*a*, *b*) he discussed the use of the probit transformation, and gave tables of probits

* Arithmetic and logarithmic probability paper can be obtained from the Codex Book Company, Inc., 74 Broadway, Norwood, Mass., U.S.A.

and weighting coefficients. He also described the application of the method to determinations of relative potency. In an appendix to the first of these papers, Fisher (1935) developed the maximum likelihood treatment for zero and 100 % values and showed the necessity for introducing the working probit at all levels of mortality. The question of whether the weights of the observations should be taken from the observed values, p, or from a provisional line remained under dispute for some time; though Bliss had advocated using the provisional line, the discussion on Irwin's (1937) paper indicated that there was still considerable difference of opinion. A full exposition of the technique, as finally derived by maximum likelihood methods, was given by Bliss in 1938.

Meanwhile psychologists apparently remained unaware of their method having been adopted by biologists and refined so as to be of greater theoretical soundness. A paper by Ferguson (1942) made use of the constant process for the analysis of data on the selection of items for mental tests; Lawley (1943, 1944) considered the mathematical aspects of the problems involved in this application and almost obtained an independent derivation of the maximum likelihood technique for the estimation of the parameters. Finney (1944c) has illustrated, on these same data, the use of the probit method in the form now familiar to biologists, in the hope that the developments which have taken place in that field since 1933 may become known amongst psychologists and psychophysicists to whose problems they could be adapted.

15. ALTERNATIVE RESPONSE CURVES

As alternatives to that based upon the normal distribution of dosage tolerances, several other relationships between the proportion of subjects showing the characteristic response and the dose have been proposed. One such sigmoid curve was given by Urban (1910) in the form

$$P = \tfrac{1}{2} + \frac{1}{\pi}\tan^{-1}(\alpha + \beta x), \tag{3.15}$$

where P is the proportion of subjects responding to a stimulus measured by x, and α and β are parameters to be determined. He developed a technique for the estimation of the parameters, exactly analogous to his technique for the $\Phi(\gamma)$ process, which could easily be made of full mathematical rigour by the method of Appendix II.

Wilson and Worcester (1943a) have suggested the equation

$$P = 1/\{1 + e^{-(\alpha + \beta x)}\},\qquad (3\cdot16)$$

which represents the logistic curve, a sigmoid often employed as a growth curve; P again is the proportion of subjects responding, x is the log dose, α and β are parameters. In later papers (Wilson and Worcester, 1943b, d; Worcester and Wilson, 1943) these authors have considered the maximum likelihood estimation of the parameters and the ED 50, and have also found formulae for the standard errors of estimation. Another paper (1943c) develops the theory on more general lines so as to include the probit transformation, the growth curve transformation, and others that might be wanted for certain purposes; this treatment is essentially the same as that of Appendix II, though the latter is of rather wider applicability. Berkson (1944) has also discussed the use of equation (3·16) as a dosage-response relationship.

None of these relationships has yet been found to compare in usefulness with the probit transformation, and none will be discussed further in the present book. It should be remembered that any one sigmoid curve can be transformed to a normal sigmoid, provided that the right dosage unit, in terms of which the tolerance distribution is normal, can be found. This unit need not always be a simple function of the dose, but in practice a logarithmic scale is very frequently good enough for the purpose and many different sigmoid curves are effectively normalized by the same choice of scale.

Chapter 4

THE MAXIMUM LIKELIHOOD SOLUTION

16. Working Probits

Though the methods described in Chapter 3 are often adequate for estimating the LD 50, the variance of tolerances, or other quantities connected with the dosage-response relationship, complications in the data may prevent these being used satisfactorily and an arithmetical rather than a graphical technique for obtaining the best fitting probit regression line is then required. For example, the observations, when plotted as in Fig. 8, may be too irregular for any confidence to be placed in an eye estimation of the fitted line, or differences in the weights attached to the observations may be so marked as to make satisfactory allowance for them very difficult without objective computational methods. The need becomes still greater when several different materials or dose factors are simultaneously under test; in the more complex types of experiment discussed in Chapters 5–8 developments of the analytical technique now to be discussed are almost essential.

The mathematical basis of the method of estimating the probit regression equation (3·2) by a process of successive approximations is given in Appendix II, but its details need not concern the non-mathematical reader. The process is begun by drawing a provisional probit line, just as described in § 10, and using this to determine the weights, nw, to be attached to each observation. The weighted regression equation of probit mortality on dosage is then computed. The distribution of observed kills, p, about their population value, P, is not symmetrical (except for $P = 0·5$), and, as shown in Appendix II, the regression equation should therefore be calculated for *working probits** rather than for the

---

* Some writers use the term 'corrected probit' instead of working probit, but this name is an unfortunate choice; the working probit is used to give a convenient simplification of otherwise complicated calculations and is in no way a correction of the empirical value. On diagrams of probit regression lines only the empirical probits should be shown, the occurrence of zero and 100 % kills being indicated by arrows, as in Fig. 9.

empirical probits obtained directly from the proportions p. The working probit is defined in equation (II, 7) as

$$y = Y + \frac{p - P}{Z}$$

$$= Y - \frac{q - Q}{Z},$$

where Y is the *expected probit* taken from the provisional line, P and Z the corresponding probability and ordinate respectively. Either of these formulae may be used, according to convenience. The quantity

$$y_{100} = Y + \frac{Q}{Z} \qquad (4 \cdot 1)$$

is known as the *maximum working probit*; when all the test subjects in a batch are killed, $p = 1$ and the corresponding empirical probit is infinite, but the maximum working probit determined from the provisional line allows this observation to play its part in determining the regression equation. Similarly

$$y_0 = Y - \frac{P}{Z}$$

is the minimum working probit, and is used when no subjects are killed. Fisher and Yates (1943, Table XI) have tabulated the maximum working probit and also the *range*, $1/Z$, for values of Y at intervals of $0 \cdot 1$. Table III of this book shows both minimum and maximum working probits and the range; the second column of this table gives minimum working probits for $Y = 1 \cdot 1 – 6 \cdot 5$, and the fourth column gives maximum working probits for $Y = 3 \cdot 5 – 8 \cdot 9$, values outside these limits being of exceedingly rare occurrence. The difference between the maximum and minimum working probits for the same Y is the range, so that when only one is tabulated the other may easily be derived.

From Table III working probits for other values of p may be calculated as in Ex. 5 below. Table IV has been prepared in this way, so as to give working probits at intervals of $0 \cdot 1$ in Y and 1% in kill for all combinations of values whose working probits lie between 0 and 10, outside which limits working probits seldom occur.* The working probit usually does not differ from the empirical probit by as much as does the expected probit, Y.

Ex. 5. *Calculation of working probits.* As an example of the use of these tables, consider the working probit corresponding to a kill of 72·3 % when the expected probit, as taken from a provisional line, is 6·2. From Table III the minimum working probit is 1·6429 and the range 5·1497. Hence $p = 0·723$ gives

$$y = 1·6429 + 0·723 \times 5·1497$$
$$= 5·366.$$

Alternatively, using the maximum working probit, 6·7926, and $q = 0·277$
$$y = 6·7926 - 0·277 \times 5·1497$$
$$= 5·366.$$

The same result may be obtained more easily from Table IV. In this table the column for $Y = 6·2$ shows working probits of 5·351 and 5·402 for 72 and 73 % kill respectively: hence, by an interpolation which may be carried out mentally,

$$y = 5·351 + \tfrac{3}{10}(5·402 - 5·351)$$
$$= 5·366.$$

Very frequently y is required only to two places of decimals, or p is reliable only to the nearest 1 %, so that y can be read directly from Table IV with little or no interpolation.

Occasionally the expected probits may need to be taken to two decimal places. Interpolation is then necessary in order to obtain the maximum or minimum working probit and the range, after which y may be calculated as above. Thus if the value of Y in this example were 6·24 the maximum working probit and the

* A useful table of working probits, at intervals of 0·1 in Y and including every value of p that can arise in a batch of twenty or less test subjects, can be obtained from the Division of Pharmacology, Food and Drug Administration, United States Department of Agriculture, Washington, D.C. A table of minimum working probits and ranges at intervals of 0·01 in Y, prepared by W. L. Stevens, can be obtained from the same address.

range would be given by taking the sum of 6/10 of the values for $Y = 6.2$ and 4/10 of the values for $Y = 6.3$. Hence y may be written, in a form suitable for machine calculation with the minimum of resetting, as

$$y = 0.6 \times 6.7926 + 0.4 \times 6.8649$$
$$- 0.277 \times (0.6 \times 5.1497 + 0.4 \times 5.8354)$$
$$= 5.319.$$

The correct value, obtained from Stevens's more detailed table, is 5·323.

17. THE REGRESSION EQUATION

The linear regression equation of the working probits on the measure of dosage, x, is an improved estimate of the dosage-response relationship. If it differs markedly from the provisional line drawn by eye, it may itself be used as a new provisional line and the process repeated. The maximum likelihood estimate is, in fact, the limit to which these estimates tend as the cycle of determining a new line with the aid of that last calculated is indefinitely repeated. With experience the first provisional line may often be drawn so accurately that only one cycle of the calculations is needed to give a satisfactory fit, though when the empirical probits are very irregular two cycles may be needed.

De Beer (1945) has developed an ingenious system of scales and nomographs for simplifying the calculations. With the aid of these, many of the expressions required as steps in the estimation of the LD 50 and its fiducial limits can be read directly from the diagram showing the provisional regression line. The results obtained are not quite the same as the maximum likelihood values, since the method does not completely distinguish between empirical and working probits, nor does it obtain fiducial limits from the exact formula (4·6) below; nevertheless it is likely to be sufficiently good for many purposes, and merits serious consideration by those who have to make many routine LD 50 estimations but who find a purely graphical method insufficiently accurate.

4-2

Ex. 6. *Arithmetical procedure in the fitting of a probit regression line*. As a simple example of the computations required in the fitting of a probit regression line by the maximum likelihood method, the data of Ex. 1 will be used again. The computations are shown in Table 6. A more detailed account of computing procedure is given in Appendix I, where another set of data is used.

TABLE 6. Maximum Likelihood Computations for Rotenone-*Macrosiphoniella sanborni* Test

x	n	r	p (%)	Empirical probit	Y	nw	y	nwx	nwy
1·01	50	44	88	6·18	6·3	16·8	6·16	16·968	103·488
0·89	49	42	86	6·08	5·8	24·6	6·05	21·894	148·830
0·71	46	24	52	5·05	5·1	29·2	5·05	20·732	147·460
0·58	48	16	33	4·56	4·6	28·8	4·56	16·704	131·328
0·41	50	6	12	3·82	3·9	20·2	3·83	8·282	77·366
						119·6		84·580	608·472

$$1/Snw = 0·008361204, \quad \bar{x} = 0·7072, \quad \bar{y} = 5·0876.$$

$Snwx^2$	$Snwxy$	$Snwy^2$
64·42700	449·5685	3177·748
59·81418	430·3057	3095·637
4·61282	19·2628	82·111
		80·440
		1·671 $= \chi^2_{[3]}$

$$b = 4·1759,$$
$$Y = 5·0876 + 4·1759\,(x - 0·7072)$$
$$= 2·134 + 4·176x.$$

The first five columns of Table 6 are repeated from Table 2. The expected probits, Y, are obtained from the provisional line of Fig. 8, and the corresponding weighting coefficients (from Table 4) are then multiplied by n and tabulated, this step being exactly as in Ex. 2. Working probits are read from Table IV, and entered in the column y. In the example the empirical probits and working probits are so nearly the same that the former could have been used instead of y without appreciably altering the conclusions, but a routine of working with y is to be preferred. Consideration of the weights, nw, indicates that two decimal places in the working probits are quite sufficient; the variances of the individual observations are $1/nw$, and thus the standard errors are of the order of $0\cdot2$, so that differences of $0\cdot01$ in working probits are of little importance.

Individual values of the products nwx, nwy, are next tabulated and added; division of the totals by Snw gives the means $\bar{x}$, $\bar{y}$. The values of nwx are multiplied by x and added to give $Snwx^2$, then multiplied by y and added to give $Snwxy$. The products of nwy with x should be summed as a check on $Snwxy$, and also, for reasons which will shortly appear, the products of $Snwy$ with y to give $Snwy^2$. The three totals are reduced respectively by $(Snwx)^2/Snw$, $(Snwx) \times (Snwy)/Snw$ and $(Snwy)^2/Snw$, to give the sums of squares and products about the mean, $Snw(x-\bar{x})^2$, $Snw(x-\bar{x})(y-\bar{y})$ and $Snw(y-\bar{y})^2$.

From equation (II, 9) the estimated regression coefficient is

$$b = Snw(x-\bar{x})(y-\bar{y})/Snw(x-\bar{x})^2$$
$$= 19\cdot2628/4\cdot61282$$
$$= 4\cdot1759,$$

and the fitted equation is a line with this slope passing through the point $(\bar{x}, \bar{y})$:

$$Y = 5\cdot0876 + 4\cdot1759(x - 0\cdot7072)$$
$$= 2\cdot134 + 4\cdot176x.$$

Using this equation, Table 7 has been obtained in the same way as was Table 3 and gives an heterogeneity χ^2 of $1\cdot62$ with 3 degrees of freedom. Since

$$\chi^2 = S\frac{n(p-P)^2}{PQ}$$
$$= Snw\left(\frac{p-P}{Z}\right)^2,$$

the computation may be made more easily as the weighted sum of squares of deviations of the working probits from the provisional line, $Snw(y-Y)^2$. This quantity is $Snw(y-\bar{y})^2$ reduced by $\{Snw(x-\bar{x})(y-\bar{y})\}^2/Snw(x-\bar{x})^2$; the calculation is shown in Table 6, and

$$\chi^2_{[3]} = 1\cdot67,$$

which differs from the χ^2 of Table 7 only on account of rounding-off errors. This 'χ^2' is not strictly a χ^2, in the recognized meaning, until the maximum likelihood solution has been closely approximated by a series of cycles of the fitting technique, but in practice no great errors are committed by referring it to the χ^2 distribution at an earlier stage.*

TABLE 7.   Comparison of Observed and Expected Mortality in Maximum Likelihood Fitting for Rotenone-*Macrosiphoniella sanborni* Test

| Log concentration (x) | Y | P | No. of insects (n) | No. affected | | Discrepancy $(r-nP)$ | $\dfrac{(r-nP)^2}{nP(1-P)}$ |
				Observed (r)	Expected (nP)		
1·01	6·352	0·9118	50	44	45·59	−1·59	0·63
0·89	5·851	0·8026	49	42	39·33	2·67	0·92
0·71	5·099	0·5398	46	24	24·83	−0·83	0·06
0·58	4·556	0·3285	48	16	15·77	0·23	0·00
0·41	3·846	0·1242	50	6	6·21	−0·21	0·01

$$\chi^2_{[3]} = 1\cdot62$$

The χ^2 test gives no evidence of heterogeneity of departure from the fitted probit line. The variances of the parameters are therefore, from equations (II. 10, 11),

$$V(\bar{y}) = 1/Snw = 0\cdot00836, \quad V(b) = 1/Snw(x-\bar{x})^2 = 0\cdot2168,$$

whence $\bar{y} = 5\cdot088 \pm 0\cdot091, \quad b = 4\cdot176 \pm 0\cdot466.$

The slope has been altered from its provisional value of 4·01 by an amount equal to about one-third of its standard error,

* The χ^2 calculated as here does not necessarily decrease as the maximum likelihood solution is approached; the process of maximizing the likelihood function is not precisely equivalent to that of minimizing χ^2, though the final χ^2 value will usually be not very different from the minimum.

and if an accurate value of b were particularly required a further cycle of computations would be desirable; the next value obtained for b is, in fact, 4·196, the alteration being only 4 % of the standard error.

On the scale of Fig. 8 the probit equation (4·1) is practically indistinguishable from the eye estimate (equation (3·3)). The estimated log LD 50 is

$$m = \bar{x} + \frac{5 - \bar{y}}{b}$$

$$= 0 \cdot 7072 - 0 \cdot 0210$$

$$= 0 \cdot 686;$$

the variance, obtained from equation (3·6), is

$$V(m) = \frac{0 \cdot 008457}{(4 \cdot 176)^2} = 0 \cdot 0004849,$$

and therefore $\qquad s_m = 0 \cdot 0220.$

The estimate of the log LD 50 obtained in Exs. 1 and 2 (0·687 ± 0·023) agrees remarkably well with the value 0·686 ± 0·022 found by the maximum likelihood method. The LD 50 is now estimated to be 4·85 ± 0·25 mg./l.

18. HETEROGENEITY

The data of Ex. 6 were selected as being particularly regular and presenting no complications in their analysis. The χ^2 test for the heterogeneity of the discrepancies between observed and expected numbers is only valid when the expected numbers are not 'small'. At the more extreme dosages tested either P or Q is often nearly zero, so that, with the usual numbers of insects exposed to the poison, either the expected number killed (nP) or the expected number surviving (nQ) is too small for χ^2 calculated in the usual manner to be referred to the distribution in Table VI. The formulae used for χ^2 in Ex. 6 may be written

$$\chi^2_{[k-2]} = S_{yy} - S^2_{xy}/S_{xx}, \qquad (4 \cdot 2)$$

where the abbreviations S_{xx}, S_{xy} and S_{yy} are used for the sums of squares and products of deviations about means $Snw(x - \bar{x})^2$, $Snw(x - \bar{x})(y - \bar{y})$ and $Snw(y - \bar{y})^2$, and k is the number of dosages tested.* Small values of nP or nQ often lead to unduly large contributions to χ^2, thus exaggerating the significance of the total.

In other applications of the χ^2 test, a value less than 5 for the expected number in any class has often been taken as a warning that the χ^2 distribution may give misleading results (Fisher, 1944, Ex. 9). There is no special virtue in the number 5; in some circumstances considerably lower expectations produce no ill effects, and in others the χ^2 distribution may be unreliable with higher expectations (Cochran, 1942). Possibly the chief danger lies in the application of the test to data in which the expectations for most classes are moderately large but one or two are very small; when all or nearly all expectations are low the disturbance of the distribution is not likely to be so serious. Further investigation into this subject is needed, and at present no more definite advice can be given than to treat as suspect any χ^2 value most of which is made up of large contributions from classes with small expectations.

TABLE 8. Table of Greatest Expected Probit giving at least nP Expected Survivors in a Batch of n

No. in batch (n)	Expected no. of survivors (nP)			
	10	5	2	1
5	—	—	5·25	5·84
10	—	5·00	5·84	6·28
20	5·00	5·67	6·28	6·64
30	5·43	5·97	6·50	6·83
40	5·67	6·15	6·64	6·96
50	5·84	6·28	6·75	7·05
100	6·28	6·64	7·05	7·33
200	6·64	6·96	7·33	7·58
1000	7·33	7·58	7·88	8·09·

The difficulty of small expectations may usually be overcome by combining the results for extreme dosages with the next highest or next lowest, so as to build up larger expectations,

* A number of writers have used A, B and C for S_{xx}, S_{xy} and S_{yy}, but the latter symbols have the advantage of giving clearer indication of their meaning.

though some sensitivity is thereby lost. Equation (4·2) cannot then be used for the calculation of χ^2, and recourse must be had to the method of Ex. 1 involving the calculation of the various expectations. The degrees of freedom for χ^2 must be reduced by the number of dosage levels lost by combination (Ex. 7). Table 8 gives a rapid guide to the 'danger levels' of the expected numbers of survivors. The table states, for example, that the expected number of survivors in a batch of fifty insects will be less than 5 when Y exceeds 6·28 and will be less than 2 when Y exceeds 6·75. Similarly, the expected number killed will be less than 5 when Y is less than 3·72 ($= 10 - 6·28$), less than 2 when Y is less than 3·25. When the batches of insects or other test organisms are themselves small (say ten or less) it may be safe to adopt a requirement of at least 2 survivors and 2 killed in each group, but for batches of thirty and upwards a standard of about 5 seems a better working rule.

Ex. 7. *Application of probit analysis to heterogeneous data.* As a second example of the technique of fitting the probit regression line, this time to less regular data, results obtained by Busvine (1938) on the toxicity of ethylene oxide to the grain beetle, *Calandra granaria*, may be considered. These data have been fully reported by Bliss (1940 b, Table VII); for the present purpose only the records referring to insects examined 1 hr. after exposure to the poison will be used.

The data are shown in the first three columns of Table 9, x being the logarithm to base 10 of the concentration of ethylene oxide in mg./100 ml. The empirical probits were plotted against x (Fig. 9) and a provisional line

$$Y = 3·06 + 7·95x$$

was drawn by eye, allowing for a value above the line at $x = 0·391$ and a value below the line at $x = 0·033$. Using this line, by application of the maximum likelihood process, the new approximation

$$Y = 2·948 + 8·600x \qquad (4·3)$$

was reached. The change in the value of the regression coefficient was sufficiently large for a second cycle of computations to seem worth while, details of which are shown in Table 9. The values of Y are obtained from equation (4·3), the working probits from Table IV; maximum and minimum working probits respectively are required at $x = 0·391$ and $x = 0·033$, and these are read either from the 100 and 0 % lines of Table IV or directly from Table III. The computations then proceed as in Ex. 6 to give the equation

$$Y = 2·928 + 8·674x. \tag{4·4}$$

As will be seen shortly, differences between (4·3) and (4·4) are negligible by comparison with the standard errors, and no further approximation is necessary.

TABLE 9.   Maximum Likelihood Computations for Ethylene Oxide-*Calandra granaria* Test (second cycle)

x	n	r	$\frac{p}{(\%)}$	Em-pirical probit	Y	nw	y	nwx	nwy
0·394	30	23	77	5·74	6·3	10·1	5·52	3·9794	55·752
0·391	30	30	100	∞	6·3	10·1	6·86	3·9491	69·286
0·362	31	29	94	6·55	6·1	12·5	6·45	4·5250	80·625
0·322	30	22	73	5·61	5·7	15·9	5·61	5·1198	89·199
0·314	26	23	88	6·18	5·6	14·5	6·06	4·5530	87·870
0·260	27	7	26	4·36	5·2	16·9	4·38	4·3940	74·022
0·225	31	12	39	4·72	4·9	19·7	4·72	4·4325	92·984
0·199	30	17	57	5·18	4·7	18·5	5·19	3·6815	96·015
0·167	31	10	32	4·53	4·4	17·3	4·54	2·8891	78·542
0·033	24	0	0	−∞	3·2	4·3	2·74	0·1419	11·782
						139·8		37·6653	736·077

$1/Snw = 0·007153076$, $\bar{x} = 0·2694$, $\bar{y} = 5·2652$.

$Snwx^2$	$Snwxy$	$Snwy^2$
11·18778	207·3360	3986·265
10·14789	198·3159	3875·603
1·03989	9·0201	110·662
		78·241
		$32·421 = \chi^2_{[8]}$

$b = 8·6741$, $Y = 2·928 + 8·674x$.

Application of formula (4·2) gives a χ^2 of 32·42, with 8 degrees of freedom, as a measure of heterogeneity, a value which, according to Table VI, is clearly significant. The validity of this test is suspect, however, since three of the expected probits exceed 6·0 and one is less than 4·0, thus giving dangerously low values to nP or nQ. The test may be modified as shown in Table 10 by grouping the expectations for the two highest concentrations with the next highest and that for the lowest with the next lowest. Equation (4·2) is then no longer applicable, and the separate contributions to χ^2 must be calculated by the

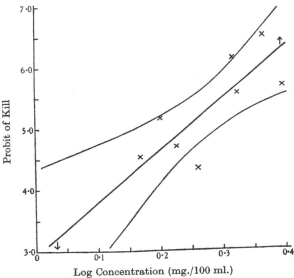

FIG. 9. Probit regression line and 5 % fiducial band for toxicity of ethylene oxide to *C. granaria* (Exs. 7–9).

method used in Ex. 1, namely, by squaring the difference between observed and expected numbers killed, multiplying by the total number, and dividing by the product of the expected kill and expected survivors. Thus

$$\frac{91 \times (0·9)^2}{81·1 \times 9·9} = 0·09$$

is the contribution from the first group. The total number of groups of insects has been reduced by 3 and therefore χ^2 has 5 degrees of freedom instead of 8;

$$\chi^2_{[5]} = 19·74$$

is still highly significant, and confirms the evidence for heterogeneity of the departures from the regression line. Since Fig. 9

gives no suggestion of any systematic deviations from the line such as would arise if the regression relationship were not truly linear, the heterogeneity may be allowed for by multiplying all variances by the heterogeneity factor (§ 11)

$$\chi^2/5 = 3\cdot95.$$

For example $V(b) = 3\cdot95/1\cdot040 = 3\cdot798,$

and therefore $b = 8\cdot674 \pm 1\cdot949.$

TABLE 10.   Comparison of Observed and Expected Mortalities for Ethylene Oxide-*Calandra granaria* Test

x	Y	P	n	r	nP	$r-nP$	$\dfrac{(r-nP)^2}{nPQ}$
0·394	6·346	0·911	30	23	27·3		
0·391	6·320	0·907	30	30	27·2	0·9	0·09
0·362	6·068	0·857	31	29	26·6		
0·322	5·721	0·765	30	22	23·0	−1·0	0·19
0·314	5·652	0·743	26	23	19·3	3·7	2·75
0·260	5·183	0·573	27	7	15·5	−8·5	10·94
0·225	4·880	0·452	31	12	14·0	−2·0	0·52
0·199	4·654	0·365	30	17	11·0	6·0	5·17
0·167	4·377	0·267	31	10	8·3		
0·033	3·214	0·037	24	0	0·9	0·8	0·08

$$\chi^2_{[5]} = 19\cdot74$$

Again, from (4·4) the log LD 50 is

$$m = 0\cdot239,$$

and using equation (3·6) once more

$$V(m) = \frac{3\cdot95}{(8\cdot674)^2}\left[0\cdot007153 + \frac{(0\cdot030)^2}{1\cdot040}\right]$$
$$= 0\cdot000421,$$

so that the standard error of m is $\pm 0\cdot021$.

If 5 % fiducial limits to m are wanted, instead of using the normal deviate, 1·96, as the multiplier of s_m, the t-value (Table VII) for 5 degrees of freedom, 2·57, must be used. Hence these limits are $0\cdot239 \pm 0\cdot054$, or 0·293 and 0·185; the LD 50 is estimated as 17·3 mg./l. with 19·6 and 15·3 mg./l. as the 5 %

fiducial limits. It will be seen later (Ex. 9) that these limits are in fact not sufficiently wide, and in this example a more exact method of estimating them is needed.

When very few degrees of freedom are available for estimating the heterogeneity factor, the corresponding t-value for any selected probability level becomes large in order to allow for the unreliability of the heterogeneity factor, and consequently the fiducial limits are widely spaced. For example, for a large number of degrees of freedom the 5 % t is about 2·00, for 5 degrees of freedom it is 2·57, for 3 degrees of freedom 3·18, and for 1 degree of freedom 12·71! This last value, in particular, is practically useless in the determination of fiducial limits. The remedy is to pool the χ^2 values from comparable series of tests, whenever this is possible, and thus to obtain a single heterogeneity factor of reasonable accuracy instead of separate factors of low precision. In particular, when a single experiment consists of several series of tests which yield similar values of b, the methods of analysis described and illustrated in § 20 should always be used; unless there are strong indications that the series differ in respect of heterogeneity, a composite test of heterogeneity is then made for the whole experiment and, if necessary, a single heterogeneity factor estimated and used for all standard errors and fiducial limits.

19. FIDUCIAL LIMITS

The expected probit, Y, for any dosage, x, has been obtained in the form

$$Y = \bar{y} + b(x - \bar{x});$$

the variance of Y is

$$V(Y) = \frac{1}{Snw} + \frac{(x - \bar{x})^2}{S_{xx}}, \qquad (4\cdot5)$$

if no allowance has to be made for heterogeneity, but the expression must be multiplied by the heterogeneity factor when this is significantly greater than unity. Fiducial limits to Y are therefore $Y \pm s_Y t$, where s_Y is the square root of $V(Y)$ and t is the normal deviate for the level of probability to be used, or, if there is significant heterogeneity, the t-value corresponding to this probability.

If the fiducial limits of Y are plotted for each x, they will be found to lie on two curves which are convex to the regression line and which approach the line most closely at the dosage $\bar{x}$. The further x is removed from $\bar{x}$ in either direction the greater is the contribution to the variance of Y from the second term of (4·5), which represents the effect of the errors of estimation of the regression coefficient b, and in consequence the more widely spaced are the fiducial limits.

In insecticidal and fungicidal investigations the effect of a single poison is often of less interest than comparisons between the effects of two or more different poisons in the same experiment. Day to day changes in the susceptibility of the organisms may alter very considerably the kill for any selected dosage without seriously upsetting the relative toxic effects of different poisons; consequently the fiducial band for one regression line may not be very helpful for inferring future behaviour (§ 20).

Ex. 8. *Fiducial bands for a probit regression line.* The plotting of fiducial limits so as to give a band on either side of the probit regression line may be illustrated on the data of Ex. 7. Under the conditions of the experiment, the true value of the kill for the range of dosages may be expected to lie within this band with a degree of confidence represented by the fiducial probability level chosen.

Formula (4·5) gives for these data

$$V(Y) = 3 \cdot 95 \times \left\{ 0 \cdot 007153 + \frac{(x - 0 \cdot 269)^2}{1 \cdot 040} \right\};$$

variances and standard errors calculated from this expression are shown in Table 11. Multiplication of the standard error, s_Y, by 2·57, the appropriate t-value for a 5 % fiducial probability, gives the width of the fiducial band on either side of the regression line; the boundaries are shown in Fig. 9.

The method of determining fiducial limits which has been described in § 12 (Ex. 3) is often sufficiently good, both for the log LD 50 and for the estimated dosage corresponding to any other selected kill; strictly speaking, however, the limits are the values of x for which the boundaries of the fiducial band attain the selected value of Y. By solving an equation so as to obtain the value of x for which Y has a selected fiducial limit, exact fiducial limits to x, the dosage giving a kill whose probit is Y, are shown to be

$$x + \frac{g}{1-g}(x - \bar{x}) \pm \frac{t}{b(1-g)} \sqrt{\left/ \left(\frac{1-g}{Snw} + \frac{(x-\bar{x})^2}{S_{xx}}\right)\right.}, \qquad (4\cdot6)$$

TABLE 11. Variance and Standard Error of Expected Probits in Equation (4·4)

x	$V(Y)$	s_Y
0·00	0·3031	0·551
0·05	0·2104	0·459
0·10	0·1367	0·370
0·15	0·0820	0·286
0·20	0·0463	0·215
0·25	0·0296	0·172
0·269	0·0283	0·168
0·30	0·0319	0·179
0·35	0·0532	0·231
0·40	0·0934	0·306

where $g = t^2/b^2 S_{xx}$ (Irwin, 1943; Fieller, 1944). Significant heterogeneity must be allowed for by increasing both g and the expression within the square root by the heterogeneity factor. When g is small compared with unity these limits are practically the same as those obtained by the method of Ex. 3, but they become more widely spaced as g approaches unity.[*]

[*] Equation (4·6) is a particular case of a very useful theorem stated by Fieller. If a and b are sample estimates of α and β subject to normally distributed random errors, and if v_{11}, v_{12}, v_{22} are joint estimates, from the same sample and based on f degrees of freedom, of the variances and covariance of a and b, then the fiducial limits of the ratio $\mu = \alpha/\beta$ are the roots of

$$\theta^2(b^2 - t^2v_{22}) - 2\theta(ab - t^2v_{12}) + (a^2 - t^2v_{11}) = 0,$$

where t is the appropriate deviate with f degrees of freedom for the chosen probability level. The limits may be written as

$$m + \frac{g}{1-g}\left(m - \frac{v_{12}}{v_{22}}\right) \pm \frac{t}{b(1-g)}\sqrt{\left[v_{11} - 2mv_{12} + m^2v_{22} - g\left(v_{11} - \frac{v_{12}^2}{v_{22}}\right)\right]}, \qquad (4\cdot7)$$

where $m = a/b$ and $g = t^2v_{22}/b^2$. In the present instance $v_{12} = 0$, but elsewhere, for example in §§ 28 and 47, this is not so.

Ex. 9. *Fiducial limits of the median lethal dose.* Reverting to the data analysed in Ex. 6, in which no heterogeneity was found, for the 5 % fiducial limits

$$t = 1 \cdot 96$$

and

$$g = (1 \cdot 96)^2/(4 \cdot 176)^2 \times 4 \cdot 613$$

$$= 0 \cdot 048.$$

The log LD 50 in this example was 0·686, and therefore the fiducial limits are, by equation (4·6),

$$0 \cdot 686 - 0 \cdot 050 \times 0 \cdot 021 \pm \frac{1 \cdot 96}{4 \cdot 176 \times 0 \cdot 952} \sqrt{\left(\frac{0 \cdot 952}{119 \cdot 6} + \frac{(0 \cdot 021)^2}{4 \cdot 613} \right)}$$

or 0·729 and 0·641, as compared with 0·729 and 0·643 obtained from the approximation $m \pm 1 \cdot 96 s_m$. With so small a value of g, the difference between the two methods is trivial.

In Ex. 7, g is much larger; taking in the heterogeneity factor, and using the 5 % value of t with 5 degrees of freedom,

$$g = 3 \cdot 95 \times (2 \cdot 57)^2/1 \cdot 040 \times (8 \cdot 674)^2$$

$$= 0 \cdot 333,$$

a value which is certainly not negligible by comparison with unity. In this example $m = 0 \cdot 239$ and the fiducial limits are therefore

$$0 \cdot 239 - 0 \cdot 499 \times 0 \cdot 030 \pm \frac{2 \cdot 57}{8 \cdot 674 \times 0 \cdot 667} \sqrt{\left[3 \cdot 95 \times \left(\frac{0 \cdot 667}{139 \cdot 8} + \frac{0 \cdot 0009}{1 \cdot 040} \right) \right]}$$

or $0 \cdot 224 \pm 0 \cdot 066$. These limits, 0·290 and 0·158, are much wider than the values of 0·293 and 0·185 obtained in Ex. 7 by the simpler method. They may be read directly from the boundaries of the fiducial band in Fig. 9. On the concentration scale the limits are 19·5 and 14·4 mg./l.

Chapter 5

THE COMPARISON OF EFFECTIVENESS

20. RELATIVE POTENCY

CHANGES in the level of tolerance of the population of test subjects frequently make it impossible to rely on assays of materials carried out singly. It is therefore customary, in many types of investigation, to assay a test material against a standard and to measure the performance in relation to that of the standard rather than as an absolute effect. In the second of his early papers on the probit method and its applications, Bliss (1935b) considered the measurement of differences between two or more comparable series of dosage mortality records. His suggestion of measuring differences in terms of relative potency has been widely adopted in the comparison of toxicity data and in biological standardization procedure. Examples have been given by Irwin (1937) and Cochran (1938).

When the dosage scale, in which tolerances of the test subjects in respect of a stimulus are normally distributed, is logarithmic the variances of the tolerances for a number of closely related stimuli are often nearly equal. For example, derris derivatives, such as rotenone, deguelin, toxicarol, and elliptone, have been found (Tattersfield and Martin, 1938; Martin, 1942) to have similar tolerance variances of their log concentrations when used against *Aphis rumicis* or against *Macrosiphoniella sanborni* under carefully controlled conditions of spraying. This equality of variances is shown in the probit analysis by parallelism of the probit regression lines; the comparison of different series of data is then particularly simple. Parallel regression lines are much less likely to be found with non-logarithmic dosage scales.

The *relative potency* of two stimuli is defined as the ratio of equally effective doses. If two series of quantal response data yield parallel probit regressions against the logarithm of the dose, there is a constant difference between dosages producing the same proportion of responding subjects, and consequently a constant relative potency at all levels of response; the relative potency then provides a convenient description of the difference between the two series. The constant dosage difference is usually denoted by M, and is equal to the difference between the median effective dosages. If suffices are used to indicate the two series, the *relative dosage value* of the first series with respect to the second, or the amount by which a dosage in the first series is less than an equally effective dosage in the second, is

$$M_{12} = m_2 - m_1 = \bar{x}_2 - \bar{x}_1 - (\bar{y}_2 - \bar{y}_1)/b, \qquad (5\cdot1)$$

b being the common slope of the two regression lines. For data having $x = \log_{10} \lambda$, the relative potency is then

$$\rho_{12} = 10^{M_{12}}. \qquad (5\cdot2)$$

The variance of M_{12} is

$$V(M_{12}) = \frac{1}{b^2} \{V(\bar{y}_1) + V(\bar{y}_2) + (\bar{x}_2 - \bar{x}_1 - M_{12})^2 \, V(b)\}; \qquad (5\cdot3)$$

the standard error, s_M, derived from this variance may be used for assigning approximate fiducial limits to M, but, unless b has been estimated with such precision that $g = t^2 V(b)/b^2$ is small, these limits will be too narrow. As in § 19, exact fiducial limits may be determined (Cochran, 1938) and take the form

$$M_{12} + \frac{g}{1-g} \, (M_{12} - \bar{x}_2 + \bar{x}_1)$$

$$\pm \frac{t}{b(1-g)} \sqrt{[(1-g)\{V(\bar{y}_1) + V(\bar{y}_2)\} + (\bar{x}_2 - \bar{x}_1 - M_{12})^2 \, V(b)]}; \qquad (5\cdot4)$$

t is the normal deviate for the significance level, unless heterogeneity requires it to be taken from Table VII with the number of degrees of freedom in χ^2, in which case g and the variances must be increased by the heterogeneity factor. This expression, whose analogy with (4·6) is apparent, reduces to $M_{12} \pm s_M t$ when g is very small.

The numerical estimation of M is accomplished by straightforward extension of the methods of Chapters 3 and 4. If inspection of the probit diagram suggests that the two regression lines

are parallel, the provisional lines are drawn parallel. Working
probits are formed and sums of squares and products calculated
for each series separately. A new approximation to the common
slope, with which the computations may be repeated if necessary,
is given by adding the components from each series, and taking

$$b = \frac{_1S_{xy} + _2S_{xy}}{_1S_{xx} + _2S_{xx}} = \frac{\Sigma S_{xy}}{\Sigma S_{xx}}, \qquad (5\cdot5)$$

where Σ indicates summation over the two series. When a satisfactory estimate of b has been obtained, the heterogeneity of
the data may be tested in the usual manner by comparing observed and expected numbers of test subjects; the most rapid
means of deriving the value of χ^2 for this test is illustrated
in Ex. 10. If there is no heterogeneity, the variance of b is

$$V(b) = 1/\Sigma S_{xx}, \qquad (5\cdot6)$$

so that $\quad V(M_{12}) = \dfrac{1}{b^2} \left\{ \dfrac{1}{_1Snw} + \dfrac{1}{_2Snw} + \dfrac{(\bar{x}_2 - \bar{x}_1 - M_{12})^2}{\Sigma S_{xx}} \right\}. \qquad (5\cdot7)$

These variances must be increased by the heterogeneity factor
if this is significantly greater than unity.

Though the relative potency of two poisons can be estimated
from data on only two doses of each, provided that the regression
lines do not markedly depart from parallelism, it is preferable
to have at least three doses in each series. In the first discussion
of relative potency in relation to probit analysis, Bliss (1935b)
has given expressions for M when only two doses have been used
for each series and also when a single dose of the test material
is assayed against a series of doses of a standard. The latter is
an unsatisfactory method of assaying relative potency, as there
can be no certainty that the regression line will be parallel to
that for the standard, yet the assumption of parallelism is im-
plicit in the estimation and interpretation of M. The method
depending on only two doses of each poison should not be used
unless the regression relationship is already known to be linear,
since the data themselves can provide no information on the
existence of a curvature on the x-scale used (Finney, 1944b).

Ex. 10. *Relative potency of rotenone, deguelin concentrate, and
a mixture of the two.* The data considered in Exs. 1, 4 and 6 are

one portion of the results of an experiment fully reported by Martin (1942) in which a deguelin concentrate and a 1 : 4 mixture of rotenone and deguelin concentrate were tested as well as rotenone alone. The complete results are shown in Table 12, and the empirical probits are plotted in Fig. 10.

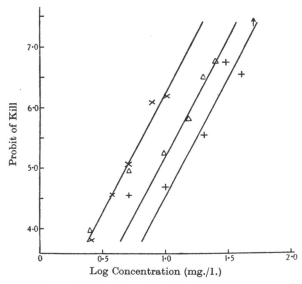

FIG. 10. Probit regression lines for estimation of relative potency of rotenone, a deguelin concentrate, and a mixture of the two (Ex. 10). × rotenone; + deguelin concentrate; △ mixture.

From the figure the lowest concentration of the deguelin concentrate and the two lowest concentrations of the mixture are seen to give points markedly disagreeing with any set of parallel straight lines. The phenomenon of a 'break' in the regression line at low concentrations has been observed in a number of toxicological investigations; here it may be the result of the poisons not being chemically pure substances (Bliss, 1939a). Dosage-response relationships for mixed poisons are discussed in Chapter 8, but for the present data the difficulty may be overcome by rejecting the three aberrant points from the analysis. Though generally undesirable, this course may be defended here on the grounds that the chief interest lies in the behaviour of

TABLE 12. Maximum Likelihood Computations for Estimation of Relative Potency of Rotenone, a Deguelin Concentrate, and a Mixture of the Two

x	n	r	p (%)	Empirical probit	Y	nw	y	nwx	nwy
				Rotenone					
1·01	50	44	88	6·18	6·3	16·8	6·16	16·968	103·488
0·89	49	42	86	6·08	5·8	24·6	6·05	21·894	148·830
0·71	46	24	52	5·05	5·1	29·2	5·05	20·732	147·460
0·58	48	16	33	4·56	4·6	28·8	4·56	16·704	131·328
0·41	50	6	12	3·82	3·9	20·2	3·83	8·282	77·366
						119·6		84·580	608·472
				Deguelin concentrate					
1·70	48	48	100	∞	7·3	3·6	7·68	6·120	27·648
1·61	50	47	94	6·55	6·9	7·7	6·42	12·397	49·434
1·48	49	47	96	6·75	6·4	14·8	6·67	21·904	98·716
1·31	48	34	71	5·55	5·8	24·1	5·53	31·571	133·273
1·00	48	18	38	4·69	4·5	27·9	4·70	27·900	131·130
0·71	49	16	33	4·56	3·4	—	—	—	—
						78·1		99·892	440·201
				Mixture					
1·40	50	48	96	6·75	6·8	9·0	6·75	12·600	60·750
1·31	46	43	93	6·48	6·4	13·9	6·47	18·209	89·933
1·18	48	38	79	5·81	5·9	22·6	5·80	26·668	131·080
1·00	46	27	59	5·23	5·2	28·9	5·23	28·900	151·147
0·71	46	22	48	4·95	4·0	—	—	—	—
0·40	47	7	15	3·96	2·8	—	—	—	—
						74·4		86·377	432·910

$$\frac{1}{_rSnw} = 0.008361204, \qquad \bar{x}_r = 0.7072, \qquad \bar{y}_r = 5.0876,$$

$$\frac{1}{_dSnw} = 0.012804097, \qquad \bar{x}_d = 1.2790, \qquad \bar{y}_d = 5.6364,$$

$$\frac{1}{_mSnw} = 0.013440860, \qquad \bar{x}_m = 1.1610, \qquad \bar{y}_m = 5.8187.$$

	$Snwx^2$	$Snwxy$	$Snwy^2$	
Rotenone	64·42700	449·56848	3177·74804	
	59·81418	430·30570	3095·63691	
	4·61282	19·26278	82·11113 −	80·440 = 1·671
Deguelin concentrate	132·03910	578·40765	2541·44933	
	127·76455	563·02891	2481·13855	
	4·27455	15·37874	60·31078 −	55·329 = 4·982
Mixture	101·86203	508·68363	2542·69182	
	100·28207	502·60036	2518·96597	
	1·57996	6·08327	23·72585 −	23·422 = 0·304
Total	10·46733	40·72479	166·14776 − 158·446 = 7·702	

the various toxic materials at high concentrations, in which region Fig. 10 suggests that the data may be adequately represented by three parallel lines.

Provisional lines for the three series of data have been drawn in the figure, and have equations:

$$Y_r = 2 \cdot 30 + 3 \cdot 94x, \quad Y_d = 0 \cdot 59 + 3 \cdot 94x, \quad Y_m = 1 \cdot 24 + 3 \cdot 94x.$$

From these, first approximations to the log LD 50's are

$$m_r = 0 \cdot 685, \quad m_d = 1 \cdot 119, \quad m_m = 0 \cdot 954.$$

A provisional estimate of the relative dosage value of rotenone and deguelin concentrate is therefore

$$M_{rd} = 1 \cdot 119 - 0 \cdot 685 = 0 \cdot 434,$$

giving a relative potency

$$\rho_{rd} = 10^{0 \cdot 434} = 2 \cdot 72.$$

Expected probits, Y, are read from the provisional lines and entered in Table 12. Calculations of sums of squares and products proceed for each series separately, those for rotenone being identical with those of Ex. 6, since the same set of provisional probits has been used. Corresponding values of S_{xx}, S_{xy} and S_{yy} are then summed over the three series; for example

$$\Sigma S_{xx} = 4 \cdot 61282 + 4 \cdot 27455 + 1 \cdot 57996$$
$$= 10 \cdot 46733.$$

These are used to give the new estimate of the regression coefficient

$$b = 40 \cdot 72479 / 10 \cdot 46733$$
$$= 3 \cdot 8907.$$

The three regression lines are lines through the three mean points, $(\bar{x}, \bar{y})$, with slope b; thus

$$\left. \begin{aligned} Y_r &= 5 \cdot 0876 + 3 \cdot 8907(x - 0 \cdot 7072) \\ &= 2 \cdot 336 + 3 \cdot 891x, \\ Y_d &= 0 \cdot 660 + 3 \cdot 891x, \\ Y_m &= 1 \cdot 302 + 3 \cdot 891x. \end{aligned} \right\} \qquad (5 \cdot 8)$$

The differences from the provisional lines are small, leaving little doubt that the hypothesis of a single value of b for the three series is justified. A test of parallelism is given by comparing the sum of the χ^2 values for the three series with that obtained by a similar process from the total sums of squares and products. For these totals

$$\Sigma S_{yy} - (\Sigma S_{xy})^2/\Sigma S_{xx} = 7\cdot702;$$

this has 10 degrees of freedom, since there are fourteen observations and four parameters have been estimated for the equations. This quantity may be subdivided into a χ^2 with

TABLE 13.   Analysis of χ^2 for Ex. 10

	D.F.	Sum of squares	Mean square
Parallelism of regressions	2	0·745	0·372
Residual heterogeneity	8	6·957	0·870
Total	10	7·702	

8 degrees of freedom, which is the sum of the χ^2 values for the three series separately, and a residual χ^2 with 2 degrees of freedom dependent on the departure from parallelism of the three lines. The analysis of χ^2 in this way is shown in Table 13. The second line of the analysis shows a χ^2 of 6·957 for heterogeneity, a value which is clearly not significant; the sum of squares for parallelism may therefore be tested as a χ^2. Since 0·745 is not significant, there is no evidence of any conflict with the hypothesis that the three lines are parallel. Had the heterogeneity χ^2 been significant, the ratio of the two mean squares (obtained by dividing the sums of squares by the degrees of freedom) would have been tested by Fisher's tables of the variance ratio (Fisher and Yates, 1943, Table V), and the mean square for heterogeneity would have been used as the heterogeneity factor in the calculation of variances.

The new estimates of the log LD 50's are, from equations (5·8),

$$m_r = 0\cdot6847, \quad m_d = 1\cdot1154, \quad m_m = 0\cdot9504,$$

corresponding to LD50 values of 4·84, 13·04 and 8·92 mg./l. respectively. The standard error of m may be calculated from equation (3·6), as in Ex. 6, remembering that b has now been estimated from the more extensive data, so that, by equation (5·6),

$$V(b) = 1/10\cdot4673.$$

The revised value of the estimate of the relative dosage value of rotenone and the deguelin concentrate is

$$M_{rd} = 0\cdot4307 \pm 0\cdot0390,$$

the standard error being derived as the square root of the variance; from equation (5·7)

$$V(M) = \frac{1}{(3\cdot8907)^2}\left\{\frac{1}{119\cdot6} + \frac{1}{78\cdot1} + \frac{(0\cdot5718 - 0\cdot4307)^2}{10\cdot4673}\right\}$$
$$= 0\cdot00152.$$

The relative potency is therefore

$$\rho_{rd} = 2\cdot70 \pm 0\cdot24,$$

the standard error of the antilogarithm of M being obtained by equation (3·7). Rotenone is thus estimated to be 2·70 times as toxic to $M.$ *sanborni* as the deguelin concentrate, or, in other words, in order to give a kill as great as that produced by a given concentration of rotenone a 2·70 times greater concentration of deguelin concentrate would be required.

For the 5 % fiducial limits of M, $t = 1\cdot960$ and therefore

$$g = 3\cdot84/10\cdot47 \times 15\cdot14$$
$$= 0\cdot024,$$

a value sufficiently small to be ignored; consequently fiducial limits of M may be taken as $M \pm 1\cdot960 \times 0\cdot390$ or 0·5071 and 0·3543. Hence the fiducial limits for ρ are 3·21 and 2·26.

In the same manner, the potency of the mixture relative to rotenone or the deguelin concentrate may be estimated. Further consideration of these three relative potencies will be deferred until Ex. 20, where the data will be used as an illustration of similar joint action of two poisons.

21. COMBINATION AND COMPARISON
OF RELATIVE POTENCIES

When the relative dosage value of two poisons has been assayed separately by two or more series of tests, an improved estimate is given by the weighted mean, using for weights the reciprocals of the variances of the estimates of M (Cochran, 1938). A weighted sum of squares of deviations of M may be used as a χ^2 in a test of the heterogeneity of the estimates averaged (Miller, Bliss and Braun, 1939). When there is no evidence of heterogeneity the sum of the weights is the weight to be attached to the mean; when a significant χ^2 occurs the total weight must be divided by the heterogeneity factor.

Ex. 11. *Combination of relative potencies.* In separate tests Tattersfield and Martin (1938, Table I) have compared a toxicarol precursor (potash separated), prepared in three different ways, with rotenone. The dosage values relative to rotenone were $-1{\cdot}164 \pm 0{\cdot}036$, $-1{\cdot}186 \pm 0{\cdot}029$ and $-1{\cdot}114 \pm 0{\cdot}042$. Differences between these barely exceed their standard errors, and it is therefore reasonable to form a combined estimate using the reciprocals of the squares of the standard errors as weights. The estimate is

$$M = -\frac{772 \times 1{\cdot}164 + 1189 \times 1{\cdot}186 + 567 \times 1{\cdot}114}{772 + 1189 + 567}$$

$$= -2940{\cdot}4/2528$$

$$= -1{\cdot}163.$$

The variance of M is $0{\cdot}000396$, the reciprocal of the sum of the weights, and the standard error is therefore $\pm 0{\cdot}020$. The test of heterogeneity of the separate estimates is scarcely required here, in view of the close agreement between them; the appropriate χ^2 is

$$\chi^2_{[2]} = 772 \times (1{\cdot}164)^2 + 1189 \times (1{\cdot}186)^2$$
$$+ 567 \times (1{\cdot}114)^2 - (2940{\cdot}4)^2/2528$$
$$= 1{\cdot}99.$$

With weights as nearly equal as these, little is lost by taking the unweighted mean

$$M' = -(1{\cdot}164 + 1{\cdot}186 + 1{\cdot}114)/3$$
$$= -1{\cdot}155,$$

the variance of which is

$$V(M') = \{(0{\cdot}036)^2 + (0{\cdot}029)^2 + (0{\cdot}042)^2\}/3^2$$
$$= 0{\cdot}000433.$$

This estimate therefore has a standard error of $\pm 0{\cdot}021$ and is only 9 % less precise than the weighted mean.

If two poisons have been tested in different experiments, even though they show similar values of b a direct estimation of their relative dosage value as the difference in the estimated log LD 50's may be misleading: the level of susceptibility of the test subjects may have been different in the two experiments or experimental conditions may have differed in such a way that the absolute potency of any poison is different in the two experiments though relative potencies within an experiment are unaffected. An estimate of relative potency independent of any difference in susceptibility may still be made, providing that each poison can be compared with a third which has been tested in both experiments. Estimates of M_{13} and M_{23} can then be made 'within experiments'; the required M is estimated as

$$M_{12} = M_{13} - M_{23}, \tag{5.9}$$

and

$$\rho_{12} = \rho_{13}/\rho_{23}.$$

The variance of M_{12} is the sum of the variances of M_{13} and M_{23}.* The method should only be used when the conditions in the two experiments are sufficiently similar for the estimate obtained to be relevant.

* Cochran (1938) suggested this procedure for testing the significance of a difference in potency of poisons used in separate experiments. The significance should be assessed from the normal distribution, not the t-distribution, unless the variances have been adjusted by a heterogeneity factor.

Ex. 12. *Comparison of relative potencies.* Tattersfield and Martin (1938, Table I), experimenting with rotenone and a resin derived from a Sumatra-type derris root, found a relative dosage value

$$M_{sr} = -0.800 \pm 0.025.$$

In a second experiment, using rotenone and a *Derris elliptica* resin, they found

$$M_{er} = -0.552 \pm 0.031.$$

Both experiments were spraying trials against *Aphis rumicis*, the same technique being used for both. Now

$$(0.025)^2 + (0.031)^2 = 0.001586 = (0.040)^2;$$

hence the relative dosage value of the two resins is estimated as

$$M_{se} = -0.248 \pm 0.040,$$

and the Sumatra-type resin is estimated to be 0·56 times as toxic as the *Derris elliptica*.

Ex. 13. *The relative toxicities of seven derris roots.* A further example is provided by earlier data from Tattersfield and Martin (1935) in which seven samples of derris root were tested for their toxicity to *Aphis rumicis*. Tests were made on six occasions, a different pair of roots being compared on each occasion. The proportions of badly affected, moribund, and dead at each concentration have been re-examined for this example, using as the measure of dosage the logarithm of the concentration of ether extract in mg./l. The log LD 50 and M values for these six tests were found to be (the suffices referring to the seven roots)[*]:

Test 1 $m_2 = 1.539,$ $m_5 = 1.607,$ $M_{25} = 0.068,$

,,   2 $m_5 = 1.695,$ $m_6 = 1.452,$ $M_{56} = -0.243,$

,,   3 $m_5 = 1.622,$ $m_7 = 1.384,$ $M_{57} = -0.238,$

,,   4 $m_4 = 1.631,$ $m_5 = 1.682,$ $M_{45} = 0.051,$

,,   5 $m_1 = 1.496,$ $m_3 = 1.492,$ $M_{13} = -0.004,$

,,   6 $m_1 = 1.440,$ $m_7 = 1.421,$ $M_{17} = -0.019.$

[*] Allowance was made for the natural death rate estimated from control batches, the methods of Chapter 6 being used.

From these estimates the relative potency of any two roots can be estimated by comparisons within experiments. For example, the relative potency of roots 1 and 4 may be estimated from the relative dosage value

$$M_{14} = M_{17} - M_{57} - M_{45} = 0.168;$$

the standard errors are not shown here, but are obtainable by the same method as before. A convenient way of summarizing the LD 50's is to take a mean value for root no. 5, which was tested four times, and to place the others at the estimated distances from this; this mean might be obtained by appropriate weighting, as in Ex. 11, but the simple arithmetic mean of the four values, 1·652, is almost equally good. The log LD 50's may then be written as:

Root no. 1: 1·433   Root no. 5: 1·652

,,   2: 1·584   ,,   6: 1·409

,,   3: 1·429   ,,   7: 1·414

,,   4: 1·601

The values are very similar to those computed independently by Bliss (1939 a). The variance of M_{14} is the sum of the variances of M_{17}, M_{57} and M_{45}, each being taken from a separate experiment, and the standard error of any relative dosage value should be obtained in this way rather than by using standard errors of each log LD 50, such as Bliss has estimated.

Had more tests of pairs of these seven roots been carried out, it would have been possible to estimate some of the relative potencies by more than one chain of pairs. In order to obtain the best possible estimates the method of least squares would then have had to be used (see Ex. 14), though data from carefully controlled trials might be sufficiently consistent for satisfactory estimates to be made, by simple averaging, without this process.

Ex. 14.   *Combination of relative potencies by the method of least squares*. As a more complex example of the estimation of relative potencies from a series of tests, data published by Martin (1940) on tests of four different derris roots as poisons for *A. rumicis* will be considered. Of these roots, known as W. 211, W. 212, W. 213 and W. 214, two sets of three were tested on different occasions, and on a third occasion W. 211 and W. 213 were compared with rotenone. The estimated log LD 50's are shown in Table 14, together with the weights (reciprocals of the variances) to be attached to each. For each occasion the three estimates were obtained from a common regression coefficient as in Ex. 10, but in the computations modified weighting coefficients (§ 27) were used in order to take account of 2–4 % mortality amongst the control batches of insects.

TABLE 14. Estimated log LD 50's and their Weights for Rotenone and Four Derris Roots on Three Occasions

Date	Rotenone	W. 211	W. 212	W. 213	W. 214	Totals
15. vi. 38	—	2·430 (1790)	2·228 (3610)	—	2·022 (4820)	22138·82 (10220)
27. vii. 38	—	2·420 (3520)	2·156 (2530)	2·255 (3290)	—	21392·03 (9340)
18. vii. 39	0·980 (1640)	2·193 (1400)	—	2·191 (1420)	—	7788·62 (4460)
	1607·20 (1640)	15938·30 (6710)	13497·76 (6140)	10530·17 (4710)	9746·04 (4820)	51319·47 (24020)

$$m = 2\cdot1365308.$$

The estimated log LD 50's for any one occasion are not independent, since they are based on the same regression coefficient. Hence the variance of a relative dosage value obtained for two poisons on one occasion is not simply the sum of two variances of the form of equation (3·6), but is an expression like equation (5·3). In this example the second term of equation (3·6) was in every instance small relative to the first and thus had little effect on the variance, as may be expected when knowledge of the poisons allows the experiment to be so planned that the mean probit, $\bar{y}$, shall be near to 5. The complication of this non-independence has therefore been ignored in the analysis of the data. In order to estimate the relative potencies of the five materials under comparable conditions, account must be taken of the difference in susceptibility of the insects on the three occasions. If the relative potency of any pair remained the same on all occasions, the $\log LD\,50$ of one material on one occasion is expressible as the sum of three components, the general mean, a root or column constant, and an occasion or row constant. For example, the $\log LD\,50$ for root W. 211 on 27 July 1938 should be expressible as $m + c_1 + r_2$, where m is the general (weighted) mean, c_1 is one of the five column constants c_r, c_1, c_2, c_3, c_4 corresponding to the root, and r_2 is one of the three row constants r_1, r_2, r_3 corresponding to the occasion. Since each set of constants is to represent deviations about the general mean, the weighted totals must be zero, so that

$$1640c_r + 6710c_1 + 6140c_2 + 4710c_3 + 4820c_4 = 0,$$
$$10220r_1 + 9340r_2 + 4460r_3 = 0. \tag{5.10}$$

Estimates of all the constants can be obtained by minimizing the weighted sum of squares of differences between the values in Table 14 and their expectations in terms of the constants. The minimizing conditions are simply statements that the weighted total for any one of the columns or rows is equal to its expected value. These weighted totals are shown in the margins of Table 14, with the total weights in brackets. Thus, for example, the second column gives the equation

$$6710m + 1790(c_1 + r_1) + 3520(c_1 + r_2) + 1400(c_1 + r_3) = 15938 \cdot 30.$$

The general mean is

$$m = 51319 \cdot 47/24020 = 2 \cdot 1365308,$$

and the equations therefore reduce to the following form, in which only the coefficients of the eight unknowns are tabulated on the left-hand side:

c_r	c_1	c_2	c_3	c_4	r_1	r_2	r_3	
1640	.	.	.	.	.	.	1640	$= -1896 \cdot 71,$
.	6710	.	.	.	1790	3520	1400	$= 1602 \cdot 18,$
.	.	6140	.	.	3610	2530	.	$= 379 \cdot 46,$
.	.	.	4710	.	.	3290	1420	$= 467 \cdot 11,$
.	.	.	.	4820	4820	.	.	$= -552 \cdot 04,$
.	1790	3610	.	4820	10220	.	.	$= 303 \cdot 48,$
.	3520	2530	3290	.	.	9340	.	$= 1436 \cdot 83,$
1640	1400	.	1420	.	.	.	4460	$= -1740 \cdot 31.$

$$(5.11)$$

These equations are not entirely independent, and, before they can be solved, need to be supplemented by the two conditions on weighted totals already stated. Though they appear complicated, a little study of their structure should enable the analogous equations for other similar sets of data to be written down quite easily. The solution is not difficult, but is tedious; the results to any required degree of accuracy may be obtained by a process of successive approximation, and are

$$c_r = -1{\cdot}0177, \qquad r_1 = 0{\cdot}0576,$$
$$c_1 = 0{\cdot}2507, \qquad r_2 = 0{\cdot}0033,$$
$$c_2 = 0{\cdot}0266, \qquad r_3 = -0{\cdot}1388.$$
$$c_3 = 0{\cdot}1387,$$
$$c_4 = -0{\cdot}1721,$$

The column constants give the required balanced comparisons between the five poisons; the row constants are of no interest except for comparing the susceptibilities of the insects on the

TABLE 15.   Comparative Potencies of Rotenone
and Four Derris Roots

Material	log LD 50	LD 50 (mg./l.)	Potency relative to W. 211	
			Present analysis	Martin (1940)
Rotenone	1·119	13·2	18·50	—
W. 211	2·387	244·0	1·00	1·00
W. 212	2·163	146·0	1·68	1·76
W. 213	2·275	188·0	1·29	1·31
W. 214	1·964	92·0	2·65	2·66

three occasions. By addition of the general mean, m, comparable values of log LD 50 for the five materials are obtained, and these are shown in the second column of Table 15. The last two columns of the table give the relative potencies as estimated by the above analysis and as previously given in Martin's paper,* where an

* When Martin's paper was written, the modified weighting coefficients (§ 27) had not been developed, and though the percentage kills were adjusted for mortality in the controls the ordinary weighting coefficients were used. As the control mortality never exceeded 4 %, the difference in method is not important.

approximate method suggested by the present writer was used; the approximation is here close to the least squares solution but might give misleading results with less regular data.

Table 16 shows the expected LD 50 values, corresponding to the experimental values in Table 14, and has been constructed by adding the appropriate pairs of row and column constants to the general mean. The goodness of the agreement of the data and the hypothesis of constant relative potencies may be assessed by means of a χ^2 test. The weighted sum of squares of deviations of the data in Table 14 is

$$1640 \times (0 \cdot 980)^2 + \ldots + 4820 \times (2 \cdot 022)^2$$

$$- (51319 \cdot 47)^2/24020 = 2779 \cdot 85,$$

and of this the fitted constants account for an amount

$$- 1896 \cdot 71 c_r + 1602 \cdot 18 c_1 + \ldots - 1740 \cdot 31 r_3 = 2765 \cdot 61.$$

TABLE 16.   Values of log LD 50 Fitted to Data of Table 14
on Assumption of Constant Relative Potencies

Date	Rotenone	W. 211	W. 212	W. 213	W. 214
15. vi. 38	—	2·445	2·221	—	2·022
27. vii. 38	—	2·390	2·166	2·278	—
18. vii. 39	0·980	2·248	—	2·136	—

Though apparently eight constants have been fitted, only six of these (two 'row' and four 'column') are independent, since the weighted sums were constrained to be zero. The residual χ^2 therefore has 2 degrees of freedom (9, less 1 for the mean and 6 for the constants); hence

$$\chi^2_{[2]} = 14 \cdot 24.$$

The same value, apart from errors of rounding off, may be obtained as the sum of squares of the discrepancies between corresponding entries in Tables 14 and 16. The significance of χ^2 indicates that the relative potencies of the five poisons did not remain constant throughout the investigation; nevertheless, the agreement between Tables 14 and 16 is probably good enough for most practical purposes.

The standard errors of the log LD 50's in Table 15 may be estimated by the use of routine processes in the method of least squares, but a detailed discussion is beyond the scope of the present work.

22. DESIGN OF EXPERIMENTS

When several poisons are to be compared in one experiment it may be impossible to test them all on one day or all on one homogeneous stock of insects. Precautions must then be taken to avoid any bias in the results arising from inequalities in the susceptibility of the insects on different days or in different stocks. A similar problem may be encountered when tests have to be made in several laboratories, or by several workers in the same laboratory; even though standardized methods of testing are used, variations may occur in the effectiveness of the same poison used by different workers. Illustrations have been given in Exs. 13 and 14 of methods of combining results from groups of tests where each group contains a different selection of poisons from all those under investigation and the general level of potency or effectiveness may vary from one group to another. These, however, are not ideal examples of how to deal with the general problem, since their lack of symmetry complicates the analysis and leads to final estimates of log LD 50's which differ widely in precision.

In planning an investigation of poisons which are too many for all to be tested on a single occasion, or which for some other reason have to be divided into groups for testing, it is usually an advantage to introduce an element of balance into the arrangement adopted, so that unbiased and reliable comparisons can be made between every pair of poisons. The levels of dose and numbers of insects should be chosen, in the light of existing knowledge, with the aim that all estimates of log LD 50 have about the same precision. Careful planning of an investigation before it is begun is always to be preferred to the haphazard accumulation of results, for no amount of detailed statistical analysis can extract satisfactory answers to the questions propounded if the experiments were badly designed for obtaining the answers.

Moore and Bliss (1942) have described an experiment in which seven different organic compounds were tested for their toxicity to *Aphis rumicis*. Sets of three compounds were tested under comparable conditions on each of seven days; the sets were so chosen that each poison was used on three days and occurred once and once only on the same day as each of the others. Moore and Bliss have discussed analysis of variance methods appropriate to these data, ignoring the differential weighting of the observations arising in the course of the usual probit analysis. They obtained unbiased comparisons between the poisons, though still more precise estimates might possibly be made from the information contained in the data. In this experiment there was a heavy mortality amongst insects sprayed with the spreader only, and in view of the considerations advanced in Chapter 6 there should be some modifications in the analysis, but the results will not be discussed here.

This experiment is said to have a *balanced incomplete block* design (Yates, 1937 b, 1940). If the seven insecticides are denoted by A, B, ..., G, the seven blocks or sets of three may be written

$$
\begin{array}{ccccccc}
A & A & A & B & B & C & C \\
B & D & F & D & E & D & E \\
C & E & G & F & G & G & F
\end{array}
$$

Yates has described how the effects of differences between days may be eliminated and the seven poisons compared by means of 'within-day' comparisons only; he has also discussed the utilization of information from comparisons 'between days', an addition which may be of value in more extensive experiments though scarcely likely to be so here. There is, of course, no advantage in this arrangement in blocks unless the blocks are likely to differ appreciably in the results which they would give for any one poison. Only experience can help the experimenter to distinguish circumstances in which the control of variations in experimental conditions by means of blocks may be expected to be advantageous from those in which it is unnecessary and may even lead to a reduction in precision. Few published accounts of researches on insecticides, or, indeed, on other aspects of biological assay, refer to this problem.

The subject of experimental design has been developed primarily in connexion with agricultural field trials, but the principles are undoubtedly useful in planning biological assays. The properties of many other designs have been studied, and valuable accounts of these have been given by Fisher (1942), Fisher and Yates (1943, Tables XV–XIX) and Yates (1937a), to whose work the reader seeking information on randomized blocks, Latin squares, confounding, and like topics must be referred; a brief discussion here would be inadequate and a full treatment would be out of place. Further research into the adaptation of the principles of experimental design and the analysis of the results of well-planned experiments is undoubtedly needed. Another branch of this subject, factorial design, is discussed in Chapter 7.

23. Precision of Assays

Miller, Bliss and Braun (1939) have discussed a number of methods for increasing the precision of estimates of M. The experimenter may always attain greater precision by using larger numbers of test subjects, but this is often inconvenient or impossible. In planning assays of M, therefore, consideration must usually be given to the most economical utilization of a limited number of subjects. If some information on the value of a relative potency already exists, doses of the test material and the standard may be chosen so as to bear this ratio to one another; the mean probits in a new assay should then be nearly equal, and the third term in the expression for the variance (equation (5·3)) will be small. In choosing the doses the experimenter should remember that very low or very high kills give low values of the weighting coefficient, w, but that, on the other hand, if the doses used are too close together the value of S_{xx} will be small and b therefore of low precision. When nothing but M is to be estimated, and some information on the LD 50's is already available, probably the best compromise is to aim at sets of expected kills lying between 4 and 6 probits and as far as possible to avoid probits lower than 3 or higher than 7. On the other hand, when little is known of the LD 50's, a sufficiently wide range of doses should be used to ensure that, at all costs, the kills obtained with each poison bracket the 50 % point. The slope of the regression lines, being the reciprocal of the standard deviation of the dosage-tolerance distribution, is also of some interest, and for its satisfactory estimation a rather wider range of doses (and consequently of kills) may be desirable than would be ideal if M alone were considered.

The expression for the variance of M is also reduced by any procedure which increases the slope of the regression lines. The regression coefficient, b, however, is not under the control of the experimenter in the same way as are the dose levels, but is a characteristic of the poison, the test subject, and the conditions of the experiment. Though the precision of an assay of relative potency may be increased by a change in the test subject or in the experimental conditions, there may then be some doubt of whether the same quantity is being estimated. Bliss and Cattell (1943, pp. 482–5) refer to instances of drug assay with animals which have given estimates of relative potency very different from the values found for man, in some cases even the order of potency being reversed. The relative potencies of insecticides may be altered by a change in spray medium or in method of application, and may be entirely different for different species of insect.

It might be thought that, in assaying the potency of a poison or other stimulus relative to a standard, precision would be increased by comparing the response to the substance under test with that found for the standard in all previous tests, rather than with the small amount of data for the standard obtained in the course of the current assay. The danger in this approach is that, though conditions can usually be so controlled that relative potency remains reasonably constant in tests made over a period of time, absolute potency often varies very considerably from day to day. The data examined in Ex. 14, for example, indicate that the poisons on the last occasion of the tests averaged 60 % higher potency than on the first. Bliss and Packard (1941) have reported that exposure of eggs of *Drosophila melanogaster* to röntgen rays gave the same probit regression relationship between kill and intensity of irradiation on a number of occasions over a period of seven years, and Whitlock and Bliss (1943) have described a series of antihelminthic tests in which the position of the regression line remained the same, within the limits of experimental error. Nevertheless, this state of affairs appears to be exceptional, and in any new investigation the experimenter should be prepared for much greater instability in the mean log tolerance, m, than in the slope, b, until experience shows otherwise.

24. Mean Probit Difference

A second measure of the difference between two series of observations giving parallel probit regressions is the *mean probit difference* (Finney, 1943a). The mean probit difference, Δ, between two parallel probit lines is defined as the constant vertical difference between the lines

$$\Delta_{12} = Y_1 - Y_2 = bM_{12}$$
$$= \bar{y}_1 - \bar{y}_2 - b(\bar{x}_1 - \bar{x}_2). \qquad (5\cdot12)$$

The variance of this expression is

$$V(\Delta_{12}) = V(\bar{y}_1) + V(\bar{y}_2) + (\bar{x}_1 - \bar{x}_2)^2 \, V(b), \qquad (5\cdot13)$$

which may be written

$$V(\Delta_{12}) = \frac{1}{{}_1Snw} + \frac{1}{{}_2Snw} + \frac{(\bar{x}_1 - \bar{x}_2)^2}{\Sigma S_{xx}}$$

when no heterogeneity factor is needed.

Though Δ is in some ways easier to use than M, particularly in respect of its fiducial limits which do not require complex expressions like (5·4), it has the serious disadvantage of giving a comparison that is much less readily interpreted in practical terms. Indeed, Δ only measures the difference in effect of equal doses—and that in a unit whose value in percentages is not constant at all parts of the scale—instead of comparing the magnitude of equally effective doses. The mean probit difference may sometimes provide a convenient statement of results, but is seldom useful without a knowledge of the regression coefficient.

Values of Δ between different pairs of poisons and over a series of experiments may be combined and compared just as were values of M in Exs. 11–14, the weights again being the reciprocals of the variances.

Ex. 15. *Mean probit difference of rotenone and deguelin concentrate.* From equations (5·8) the mean probit difference between rotenone and the deguelin concentrate is

$$\Delta_{rd} = 1\cdot 676 \pm 0\cdot 229,$$

the standard error being obtained by equation (5·13) as

$$V(\Delta) = \frac{1}{119\cdot 6} + \frac{1}{78\cdot 1} + \frac{(0\cdot 5718)^2}{10\cdot 4673}$$
$$= 0\cdot 0524.$$

Fiducial limits for Δ are obtained by addition and subtraction of the appropriate multiple of the standard error. For rotenone and the deguelin concentrate the 5 % fiducial limits differ from Δ by $\pm\, 1\cdot 960 \times 0\cdot 229$, and are therefore 2·125 and 1·227.

25. UNEQUAL TOLERANCE VARIANCES

When the probit regression lines of two series of tests are not parallel, the interpretation of their comparative effects is more difficult. The relative dosage value of the two poisons can still be measured at a selected level of kill, but this quantity will be different at another level. The variance of such a relative dosage value is the sum of the variances of the two dosages estimated to give that kill, each being obtained from equation (3·6). Similarly, the probit difference between the poisons must be quoted for a specified dosage. Thus much of the usefulness and simplicity of these two measures is lost when the variances of the distribution of tolerance are unequal.

An apparent lack of parallelism may sometimes disappear if a more suitable x-scale is chosen, though this is unlikely unless the separate series also give indications of departing from linearity of regression on the scale first used. Up to the present no simple method of comparing the effects of two poisons whose regression lines are not parallel has been developed, though, as will appear in § 38, consideration has been given to the form of dose-mortality relationship for mixtures of two such poisons. Indeed, for two substances of totally different chemical constitutions or modes of action on the test subjects, there is no reason to expect an easy expression of the difference in the dose-response relationships and the term 'relative potency' may cease to have much useful meaning.

Chapter 6

ADJUSTMENTS FOR NATURAL MORTALITY

26. Abbott's Formula

THE responses of the test subjects have so far been assumed to be entirely due to the effects of the stimuli applied, and no allowance has been made for any responses which might have occurred without these stimuli. In some instances the assumption may be justifiable, and, indeed, almost all the numerical examples used in earlier chapters were chosen as containing evidence of little or no natural mortality under the conditions of the test. This state of affairs does not always obtain. Control batches of insects, untreated or possibly treated with a spray medium having no toxic content, often show appreciable death rates in the period between the application of insecticides to the test batches and the examination of the results of this application. A similar situation arises in ovicidal tests when it is impossible to distinguish between fertilized and unfertilized eggs at the start; the observed percentage of eggs failing to hatch must be adjusted in order to allow for the percentage unfertilized in the population from which the test batches were taken. Again, in tests of fungicides by means of spore germination counts the adjustment is required in order to allow for the existence of spores which would not germinate even in the absence of any fungicide.

If in a toxicity test a proportion C of test subjects would die even without any poison, the total death rate expected from a dose sufficient to kill a proportion P of those which would otherwise survive is

$$P' = C + P(1 - C), \tag{6.1}$$

providing that the two types of mortality operate independently. From this equation it follows that, if the total proportion dead is P', the proportion killed by the poison alone is

$$P = (P' - C)/(1 - C). \tag{6.2}$$

This is commonly known as 'Abbott's formula', on account of its use in a paper by W. S. Abbott (1925) on the adjustment of the results of insecticidal tests; in fact, it had been used earlier, for the same purpose, by Tattersfield and Morris (1924), and is an application of the well-known rule for the combination of independent probabilities.

27. APPROXIMATE ESTIMATION OF THE PARAMETERS

The effects of the adjustment for natural mortality on the maximum likelihood estimation of the parameters of the tolerance distribution have usually been ignored. Most important of these is that the effective number of subjects exposed to the poison is no longer n, the total number in a batch, but, on the average, $n(1 - C)$. This was realized by Bliss, who, in the discussion of an ovicidal test (1939a, p. 602), states: 'Both the number of eggs exposed and the percentage kill have been corrected for mortality in the untreated controls.' Examination of his correction shows the number of eggs tested at each concentration to have been reduced by the percentage mortality in the controls; thus the weights attached to each observation were proportionately reduced. The estimates of potency are not altered, but their precision is less than if no adjustment had been required.

Even when C is known exactly, this is not the only alteration required in the probit analysis. In the expression for the weighting coefficient (equation (3·5)), the product PQ arises as the variance of a binomial frequency distribution; when C is not zero, the relevant distribution is that defined by the total proportions of dead and surviving, P' and Q'. Finney (1944a) has shown that the two adjustments may be combined by taking as the weighting coefficient

$$w = \frac{Z^2}{Q\left(P + \dfrac{C}{1 - C}\right)} \tag{6·3}$$

instead of the usual $w = Z^2/PQ$, and multiplying by the unadjusted n. Values of this quantity, at intervals of 1 % in C from zero to 40 % and at intervals of 0·1 in the expected probit, are given in Table II.

When $C = 0$, (6·3) reduces to the formula for the ordinary weighting coefficients; for any other value of C these coefficients must be multiplied by $\dot{P}\Big/\Big(P + \dfrac{C}{1-C}\Big)$. Even when C is no greater than 5 %, the reduction in the value of w may be considerable, especially if the expected probit is small. Except when $C = 0$, w is not symmetrical about the value $Y = 5$, but decreases much more rapidly for small values of Y than for large. For most practical purposes, it is sufficient to determine w from a value of C given to the nearest 1 %, so that interpolation in Table II is seldom needed. Beyond the range of Table II, w may be calculated with the aid of Table I, as illustrated in Ex. 16.

Ex. 16. *Calculation of weighting coefficients.* Suppose that it is required to find the value of w for an expected probit $Y = 6·2$ and a control mortality of 59 %. The mortality P, corresponding to Y, must first be read from Table I as

$$P = 0·8849.$$

Also $$C/(1 - C) = 1·4390,$$

and therefore $$P\Big/\Big(P + \frac{C}{1-C}\Big) = 0·3808.$$

When $C = 0$, the weighting factor for this Y is $0·37031$ (Table II), and therefore for $C = 0·59$

$$w = 0·37031 \times 0·3808$$

$$= 0·1410.$$

This may be compared with the value for $Y = 3·8$, for which

$$P = 0·1151, \quad P\Big/\Big(P + \frac{C}{1-C}\Big) = 0·0741, \quad \text{and} \quad w = 0·0274.$$

In practice C is seldom, if ever, known exactly, and must instead be estimated from a sample of the population just as are other mortality rates. If this control batch is large relative to the batches used for the different doses in the experiment, C may be estimated satisfactorily from it alone, a procedure which ignores the information on C contributed by the mortalities observed for the test doses. The observed mortality amongst the controls, c, is itself subject to sampling variation, and, though it is an unbiased estimate of C, a better estimate may often be obtained by the use of additional information, especially that from the lower dose levels of the poisons under test. Such a dose may, for example, give a total mortality, p', which is less than c, and thus suggest that c is an overestimate of the true natural mortality, C. Similarly, a series of values of p' at small doses which are nearly equal to one another but much in excess of c, indicates that c is an underestimate. In such circumstances an improved estimate of C may be obtained either from inspection of the data or from a freehand sketch of the sigmoid response curve relating dosage to percentage kill.

When C is not too large (say less than 20 %), a probit analysis based on the estimate c, or on an estimate modified as suggested in the last paragraph, and using equation (6·3) for the weighting coefficient but otherwise proceeding as in Chapter 4, is very often sufficiently near to the maximum likelihood treatment of the data for practical purposes. The estimate of C may be greater than certain of the p', so that Abbott's formula yields negative values for p. Though no meaning can be attached to negative mortalities, the correct procedure is still to calculate and use working probits according to the rules of Chapter 4, as only in this way can each observation exercise its right influence in the estimation of the regression line or lines. Working probits corresponding to negative values of p are not often required, and have not been included in Table IV; when needed they must be calculated from Table III. No example of the use of these modified weighting coefficients in the manner just described need be given, as the method is exactly the same as that customarily employed when $C = 0$.

28. THE MAXIMUM LIKELIHOOD ESTIMATES

If a series of tests provides evidence that the natural mortality rate is high, or if the control batch is so small or the test mortalities so irregular that estimation of C is difficult, the full maximum likelihood process, following the lines indicated in

Appendix II, should be carried out. The maximum likelihood equations for the estimation of the parameters may most easily be used in the form:

$$\left.\begin{aligned}
\delta C\left\{\frac{n_c}{C(1-C)}+\frac{1}{(1-C)^2}Snw\frac{Q^2}{Z^2}\right\}+a'\frac{1}{1-C}Snw\frac{Q}{Z}+b\frac{1}{1-C}Snw(x-\bar{x})\frac{Q}{Z} \\
=\frac{n_c(c-C)}{C(1-C)}+\frac{1}{1-C}Snwy\frac{Q}{Z} \\
\delta C\frac{1}{1-C}Snw\frac{Q}{Z}\qquad\qquad +a'Snw\qquad\qquad =Snwy \\
\delta C\frac{1}{1-C}Snw(x-\bar{x})\frac{Q}{Z}\qquad\qquad +bSnw(x-\bar{x})^2 \\
=Snw(x-\bar{x})(y-\bar{y})
\end{aligned}\right\}$$

$$(6\cdot4)$$

These equations are constructed by taking a first approximation to C in the manner described in § 27, and using it in Abbott's formula to give values of p; from these, proceeding in the normal way, a provisional probit line and sets of weights and working probits are obtained. The quantity Q/Z is read from Table II, as also are the weighting coefficients, whence the coefficients in equations $(6\cdot4)$ may be rapidly calculated. The equations are then solved for δC, a', b; $C+\delta C$ is an improved estimate of the natural mortality rate, which should be used if it is decided to carry out a second cycle of the computations, and a', b are the parameters of the regression equation

$$Y = a' + b(x - \bar{x}).$$

If the revised values of the parameters are not sufficiently different from the provisional values for a new cycle of computations to be needed, a test of heterogeneity may be made by comparing the observed numbers dead and surviving at the different concentrations with the expected numbers calculated from these parameters. An example of these processes in their simplest form has been given elsewhere (Finney, 1944a), and need not be repeated here; Ex. 17 below illustrates the extension to two series of data with parallel probit regression equations.

In the general solution of equations $(6\cdot4)$, a' is no longer the weighted mean probit, $\bar{y}$. If the cycle is repeated sufficiently often, however, the maximum likelihood estimate of C is approached; this makes δC zero, so that $a' = \bar{y}$ once more and b is again simply the regression coefficient of y on x. Herein lies the justification for the approximate method of analysis described in § 27, for if a value of C sufficiently close to the maximum likelihood estimate can be chosen by inspection of the data, only the modification of the weighting coefficients is needed in order that the parameters of the regression equation may be estimated by the usual technique.

The most useful method of solving these equations is by first deriving the inverse matrix of the coefficients, or the set of c-multipliers, as described by Fisher (1944, § 29) in another connexion. From these quantities not only may the estimates of the parameters be obtained but also the standard errors of the parameters and of other measures obtained from them.

Ex. 17. *Numerical example of the maximum likelihood solution.* Martin (1940, Table 5) has reported the results of tests of the toxicity of two derris roots, W. 213 and W. 214, to the grain beetle, *Oryzaephilus surinamensis*; in this experiment over 16 % of control insects, sprayed with the alcohol-sulphonated lorol medium but no derris, were affected. The first stages of the calculations required for the maximum likelihood estimation of parameters are shown in Table 17. The logarithms of the concentrations of the two roots, measured in milligrams of dry root per litre, are shown in the first column of the table. The columns n, r, p' contain the number of insects exposed, the number affected, and the percentage affected respectively. Amongst the 129 insects in the control batch, 21 were affected, so that c is 16·3 %. The kills by the spray were all substantially higher than 16·3 %, and they appear to give little additional information on the true value of the control rate, C.

As a start to the computations a provisional value of C, which turns out to be a remarkably good guess, may be taken as 17·0 %. The percentage kills due to the poisons alone are then estimated by means of equation (6·2); for example, for the lowest concentration of root W. 213,

$$p = (0\cdot460 - 0\cdot170)/(1 - 0\cdot170) = 0\cdot349.$$

TABLE 17. Toxicity of Derris Roots W. 213 and W. 214 to *Oryzaephilus surinamensis*

x	n	r	p'	$p(C=17)$	Empirical probit	Y	nw	y	Q/Z	nwx	nwy	$nw(Q/Z)$	Equations (6·8)
								Root W. 213					
2·17	142	142	100·0	100·0	∞	7·6	4·7	7·94	0·343	10·199	37·318	1·6121	7·636
2·00	127	126	99·2	99·0	7·33	7·2	9·7	7·31	0·392	19·400	70·907	3·8024	7·160
1·68	128	115	89·8	87·7	6·16	6·3	35·0	6·15	0·565	58·800	215·250	19·7750	6·264
1·08	126	58	46·0	34·9	4·61	4·6	47·5	4·61	1·780	51·300	218·975	84·5500	4·585
							96·9			139·699	542·450	109·7395	
								Root W. 214					
1·79	125	125	100·0	100·0	∞	7·2	9·5	7·50	0·392	17·005	72·105	3·7240	7·251
1·66	117	115	98·3	98·0	7·05	6·9	14·9	7·03	0·438	24·734	104·747	6·5262	6·887
1·49	127	114	89·8	87·7	6·16	6·4	31·4	6·12	0·539	46·786	192·168	16·9246	6·412
1·17	51	40	78·4	74·0	5·64	5·5	22·9	5·64	0·876	26·793	129·156	20·0604	5·516
0·57	132	37	28·0	13·3	3·89	3·8	17·6	3·89	4·557	10·032	68·464	80·2032	3·836
							96·3			125·350	566·640	127·4384	
Control	129	21	16·3	—	—	—	—	—	—	—	—	—	—

W. 213: $\bar{x} = 1·4417$, $Snwx(Q/Z)/(1-C) = 132·22.$

W. 214: $\bar{x} = 1·3017$, $Snw(Q/Z)/(1-C) = 153·54.$

$n_c/C(1-C) = 914·25$, $n_c(c-C)/C(1-C) = -6·400.$

	$Snwx^2$	$Snwxy$	$Snwx(Q/Z)$	$Snwy(Q/Z)$	$Snw(Q^2/Z^2)$
W. 213	215·11983	820·90706	135·6391	551·9874	163·715
	201·40155	782·04048	158·2095		
	13·71828	38·86658	−22·5704		
W. 214	178·27458	779·41529	111·9036	602·8540	396·500
	163·16327	737·57346	165·8817		
	15·11131	41·84183	−53·9781		
Total	28·82959	80·70841	0·83)−76·5485	0·83)1154·8414	0·6889)560·215
			−92·227	1391·375	813·20
				−6·400	914·25
				1384·975	1727·45

The empirical probits of p are plotted, and two parallel provisional regression lines drawn in the usual manner so as to give two series of expected probits, Y. The weighting coefficients, read from Table II in the column for 17 %, are multiplied by n and the products entered in the column nw; thus, for the first line of Table 17, $Y = 7\cdot6$, the weighting coefficient is $0\cdot03298$, and this multiplied by 142 gives the weight as $4\cdot7$.* The working probits, y, are obtained in the standard manner described in § 16, and Q/Z is read from Table II for the appropriate values of Y.

Product columns, nwx, nwy and $nw(Q/Z)$, are next formed. These are multiplied in turn by x, y and Q/Z, and summed for each root separately to give the totals shown in the lower part of Table 17. The expressions $Snwx^2$, $Snwxy$ and $Snwx(Q/Z)$ are then adjusted so as to give the sums of squares and products of deviations about means; for example, the last of these is reduced by $(Snwx)(Snw\,Q/Z)/Snw$. The contributions from the two roots are added, the results divided by $(1 - C) = 0\cdot83$ or $(1 - C)^2 = 0\cdot6889$ where necessary, and other additions made in order to give the coefficients required for equations of the type of $(6\cdot4)$. Since two poisons were investigated, equations $(6\cdot4)$ have to be extended so as to give a_1', a_2' (referring to the two poisons) as well as common values of δC and b. Some coefficients have therefore to be taken from the data for the two roots separately, others from the combined values, but the method of construction of equations $(6\cdot5)$ should be made clear by a comparison with equations $(6\cdot4)$ and Table 17.

The equations are

$$\left.\begin{array}{l} 1727 \quad \delta C + 132\cdot2a_1' + 153\cdot5a_2' - 92\cdot23b \;\; = 1385\cdot0, \\ \;\; 132\cdot2 \;\; \delta C + \;\; 96\cdot9a_1' = 542\cdot45, \\ \;\; 153\cdot5 \;\; \delta C + \;\; 96\cdot3a_2' = 566\cdot64, \\ \; -92\cdot23\delta C + 28\cdot8296b = 80\cdot7084. \end{array}\right\} \quad (6\cdot5)$$

* For values of C exceeding $0\cdot40$ it is advisable also to tabulate P, the probability corresponding to Y (read inversely from Table I). The weighting coefficient may then be calculated as in Ex. 16; Finney (1944a) has illustrated this procedure.

Replacing the right-hand sides of these four equations by 1, 0, 0, 0 and solving, the first line of V, the matrix of variances and covariances, is obtained; similarly, inserting in turn on the right-hand side the sets of values 0, 1, 0, 0; 0, 0, 1, 0; and 0, 0, 0, 1, the remaining three lines are obtained. The result is

$$V = \begin{pmatrix} 0 \cdot 00099314 & -0 \cdot 0013549 & -0 \cdot 0015830 & 0 \cdot 0031772 \\ -0 \cdot 0013549 & 0 \cdot 012168 & 0 \cdot 002160 & -0 \cdot 004335 \\ -0 \cdot 0015830 & 0 \cdot 002160 & 0 \cdot 012908 & -0 \cdot 005064 \\ 0 \cdot 0031772 & -0 \cdot 004335 & -0 \cdot 005064 & 0 \cdot 044851 \end{pmatrix}$$

$$(6 \cdot 6)$$

The solution of equations (6·5) is then derived by adding the products of their right-hand sides with each of the four rows of V; thus

$$\left. \begin{aligned} \delta C &= 1385 \cdot 0 \times 0 \cdot 00099314 - 542 \cdot 45 \times 0 \cdot 0013549 \\ &\quad - 566 \cdot 64 \times 0 \cdot 0015830 + 80 \cdot 7084 \times 0 \cdot 0031772 \\ &= -0 \cdot 00003, \\ a_1' &= 5 \cdot 5981, \\ a_2' &= 5 \cdot 8847, \\ b &= 2 \cdot 7993. \end{aligned} \right\} \quad (6 \cdot 7)$$

The new estimate of the control mortality rate is $C + \delta C = 0 \cdot 16997$, a value extraordinarily close to the first approximation of 0·17. The two regression equations are

$$\left. \begin{aligned} Y_1 &= 5 \cdot 5981 + 2 \cdot 7993(x - 1 \cdot 4417) \\ &= 1 \cdot 562 + 2 \cdot 799x, \\ Y_2 &= 2 \cdot 241 + 2 \cdot 799x; \end{aligned} \right\} \quad (6 \cdot 8)$$

from them the last column of Table 17 has been calculated. The close agreement with the earlier column of expected probits is evidence that no further cycle of computations is needed, and, indeed, the agreement with the empirical probits is so good as to leave little fear of any heterogeneity of the data about the regression lines. The method of testing heterogeneity is to calculate expected numbers killed and surviving at each dose, and

from these to derive a χ^2; details of such a test, modified in order
to take account of small expectations, are given in Table 18.
For each dose level the probability P corresponding to the probit
in the last column of Table 17 is read from Table I, and, using
$C = 0\cdot170$ in equation (6·1), P' is calculated. The expected
numbers affected at each concentration are then nP', and these
are tabulated alongside r, the observed numbers. As explained

TABLE 18. Comparison of Observed and Expected Mortalities
in Derris-*Oryzaephilus surinamensis* Test

Dry root (mg./l.)	P	P'	n	r	nP'	$r - nP'$	$\dfrac{(r-nP')^2}{nP'Q'}$
			W. 213				
1480	0·9958	0·997	142	142	141·6 ⎫		
1000	0·9846	0·987	127	126	125·3 ⎬	$-0\cdot9$	0·06
480	0·8969	0·914	128	115	117·0 ⎭		
120	0·3391	0·451	126	58	56·8	1·2	0·05
			W. 214				
619	0·9878	0·990	125	125	123·8 ⎫		
458	0·9704	0·975	117	115	114·1 ⎬	$-2\cdot5$	0·52
310	0·9210	0·934	127	114	118·6 ⎭		
149	0·6971	0·749	51	40	38·2	1·8	0·34
37·1	0·1222	0·271	132	37	35·8	1·2	0·06
Controls	0	0·170	129	21	21·9	$-0\cdot9$	0·04
						$\chi^2_{[2]} = 1\cdot07$	

in Ex. 7, concentrations in which either the expected number of
killed or the expected number of survivors is less than five are
grouped; contributions to χ^2 are then calculated as in that
example, multiplying the square of the discrepancy by n and
dividing by the product of the expected numbers killed and surviving. Six groups are shown in Table 18, and four parameters,
C, a_1', a_2' and b, have been estimated from the data, so that
2 degrees of freedom remain. The value $\chi^2_{[2]} = 1\cdot07$ is not significant.

The low value of χ^2 just found is here sufficient confirmation
that the tacit assumption of a common value of b for the two
roots was justified. If the issue were in greater doubt, the test
would be to recalculate the parameters from a fresh set of
equations, allowing for different values, b_1 and b_2, and to compare

the new heterogeneity χ^2 with that in Table 18, there being one less degree of freedom when an additional parameter has been estimated. In the usual manner, the difference is a $\chi^2_{[1]}$ testing the parallelism of the regression lines. It is taken for granted that the same control mortality, C, applies to both roots, as otherwise the whole basis of the comparison is destroyed. This test is not needed here as $\chi^2_{[2]} = 1 \cdot 07$ cannot contain a component with 1 degree of freedom representing a significant departure from parallelism.

No heterogeneity factor is required for the variances and covariances of the parameters, which are the elements of the matrix V. For example, the first element, $0 \cdot 000993$, is the variance of δC, and the revised estimate of C is therefore $0 \cdot 1700 \pm 0 \cdot 0315$. As was expected in this example, the treated batches have contributed little to the precision of estimation of C; the figure from the controls alone is $c = 0 \cdot 1628$ with variance $c(1-c)/n_c$ and consequently a standard error $\pm 0 \cdot 0325$. The other entries on the diagonal of V are the variances of a'_1, a'_2 and b respectively; hence, for example, $b = 2 \cdot 799 \pm 0 \cdot 212$. The non-diagonal entries are the covariances of pairs of parameters, and are required in estimating the standard errors of quantities such as $\varDelta$, m or M.

The mean probit difference (§ 24) between the two roots is estimated to be
$$\varDelta = a'_1 - a'_2 - b(\bar{x}_1 - \bar{x}_2),$$
the variance of which is

$$
\left.
\begin{aligned}
V(\varDelta) &= V(a'_1) + V(a'_2) - 2\,\mathrm{Cov}\,(a'_1,\,a'_2) + (\bar{x}_2 - \bar{x}_1)^2\,V(b) \\
&\quad + 2(\bar{x}_2 - \bar{x}_1)\,\mathrm{Cov}\,(a'_1,\,b) - 2(\bar{x}_2 - \bar{x}_1)\,\mathrm{Cov}\,(a'_2,\,b) \\
&= 0 \cdot 0122 + 0 \cdot 0129 - 2 \times 0 \cdot 0022 + 0 \cdot 0196 \times 0 \cdot 0449 \\
&\quad + 0 \cdot 2800 \times 0 \cdot 0043 - 0 \cdot 2800 \times 0 \cdot 0051 \\
&= 0 \cdot 0214.
\end{aligned}
\right\} \quad (6 \cdot 9)
$$

Hence from equations $(6 \cdot 8)$
$$\varDelta = -0 \cdot 679 \pm 0 \cdot 146.$$

The log LD 50 for either root is given by the usual formula, with a' replacing $\bar{y}$,
$$m = \bar{x} + (5 - a')/b,$$

but equation (3·6) for the variance of m must be modified to allow for the covariance of a' and b, so that

$$V(m) = \frac{1}{b^2}\{V(a') + 2(m - \bar{x})\operatorname{Cov}(a', b) + (m - \bar{x})^2 V(b)\}. \quad (6\cdot10)$$

For root W. 213 this becomes

$$V(m_1) = (0\cdot0122 + 0\cdot0019 + 0\cdot0021)/7\cdot834$$
$$= 0\cdot00207.$$

Therefore $m_1 = 1\cdot228 \pm 0\cdot045$ and similarly $m_2 = 0\cdot986 \pm 0\cdot051$. The logarithm of the relative potency of the two roots is given by equation (5·1), and its variance is an extension of equation (5·3) to allow for the covariances between the parameters, namely,

$$\begin{aligned}V(M) = \frac{1}{b^2}\{&V(a_1') + V(a_2') - 2\operatorname{Cov}(a_1', a_2') + (\bar{x}_2 - \bar{x}_1 - M)^2 V(b)\\ &+ 2(\bar{x}_2 - \bar{x}_1 - M)\operatorname{Cov}(a_1', b) - 2(\bar{x}_2 - \bar{x}_1 - M)\operatorname{Cov}(a_2', b)\}\\ = \frac{0\cdot0122 + 0\cdot0129 - 0\cdot0043 + 0\cdot0005 - 0\cdot0009 + 0\cdot0010}{7\cdot834}\\ = 0\cdot00273.\end{aligned}$$

Hence $M = -0\cdot242 \pm 0\cdot052$. This may be compared with the result given by Martin, in his discussion of the same data, $M = -0\cdot248 \pm 0\cdot042$. The difference in the estimate of relative potency is trivial, but the present analysis shows a higher standard error since the previous overestimation of the weighting coefficients has been corrected (by a technique which was not available when Martin published his paper).

For these results $g = 0\cdot022$, a value small enough for fiducial limits to m and M to be safely calculated directly from the standard errors. Exact fiducial limits may be calculated from formula (4·7); the relative dosage value may be written

$$M = \bar{x}_2 - \bar{x}_1 - (a_2' - a_1')/b,$$

and the fiducial limits to $(a_2' - a_1')/b$ may then be found from (4·7) and (6·6) by the use of

$$V(a_2' - a_1') = V(a_2') - 2\operatorname{Cov}(a_1', a_2') + V(a_1'),$$

and $\quad \operatorname{Cov}(a_2' - a_1', b) = \operatorname{Cov}(a_2', b) - \operatorname{Cov}(a_1', b).$

Chapter 7

FACTORIAL EXPERIMENTS

29. REASONS FOR FACTORIAL DESIGN

THE introduction of the factorial principle into the planning of biological experimentation has been a revolutionary step which can now be seen as not merely useful, but essential for a full exploration of the causes underlying even the simplest biological phenomena. Fisher (1942, § 37), in an excellent chapter on the advantages of factorial experimentation, succinctly states the case for factorial design: 'In expositions of the scientific use of experimentation it is frequent to find an excessive stress laid on the importance of varying the essential conditions *only one at a time*.... This ideal doctrine seems to be more nearly related to expositions of elementary physical theory than to laboratory practice in any branch of research. In experiments merely designed to illustrate or demonstrate simple laws, connecting cause and effect, the relationships of which with the laws relating to other causes are already known, it provides a means by which the student may apprehend the relationship, with which he is to familiarise himself, in as simple a manner as possible. By contrast, in the state of knowledge or ignorance in which genuine research, intended to advance knowledge, has to be carried on, this simple formula is not very helpful. We are usually ignorant which, out of innumerable possible factors, may prove ultimately to be the most important, though we may have strong presuppositions that some few of them are particularly worthy of study. We have usually no knowledge that any one factor will exert its effects independently of all others that can be varied, or that its effects are particularly simply related to variation in these other factors. On the contrary, when factors are chosen for investigation, it is not because we anticipate that the laws of nature can be expressed with any particular simplicity in terms of these variables, but because they are variables which can be controlled or measured with comparative ease. If the investigator, in these circumstances, confines his attention to any single factor we may infer either that he is the unfortunate victim of a doctrinaire theory as to how experimentation should proceed, or that the time, material or equipment at his disposal are too limited to allow him to give attention to more than one narrow aspect of his problem.'

Factorial design has until now been most widely employed for agricultural field trials, but its value in the logical structure and interpretation of experiments is as great in the laboratory as in the field. Problems of toxicology have been investigated in the past chiefly by varying the level of a single factor in the set of conditions defining the stimulus, other factors being held as nearly constant as was practicable. Factorial design, on the other hand, entails, first, a selection of the more important factors relating to the stimulus or the subject; secondly, the adoption for the experiment of a convenient number of states or levels of each factor selected; and thirdly, the making of tests on batches of subjects under the conditions defined by various combinations of levels of these factors, non-experimental factors being held as nearly constant as possible. In this way the virtues of carefully standardized conditions are combined with the obtaining of information on the effects of variations in these conditions. The measurements of all factors constituting the stimulus may be referred to collectively as the *dose*.

In considering the desirability of adopting a factorial set of treatments, the different needs of an assay and an investigation into the laws determining the reaction of the subject to the stimulus should be borne in mind. The purpose of an assay is to assess the value of an arbitrary unit of the stimulus under test in terms of units of a standard stimulus; providing that the test stimulus can be fully described in these standard units, there will generally be no advantage in using several factors for the assay rather than only one. If, for example, an insecticidal spray whose only toxic constituent is rotenone is to be assayed in units of a standard rotenone preparation, the same result should be obtained (within the limits of sampling variation) from tests at different concentrations as from tests with different quantities of a fixed concentration, and the inclusion of different levels of other factors would have little advantage. Indeed, any indication that the result of the assay depended on other conditions (such as temperature or method of spraying) would contradict a basic assumption of the assay, and would suggest that the spray under test contained a toxic constituent whose potency could not be described simply in units of rotenone, since its equivalent in rotenone varied when experimental conditions were changed. The choice between varying the concentration and varying the quantity of spray would depend upon practical convenience and experience of which gave the more reliable results.

On the other hand, when the relationship between the reaction of the subject and the measures of the stimulus is the object of study, a factorial experiment may have many advantages. Before the action of an insecticide can be fully understood, the direct effects of various factors defining the dose and its method of application and the interactions between these must be investigated in detail. By comparison with agricultural experiments, laboratory tests of insecticides take only a short time to carry out, and, at least in the preliminary stages of a research project, a series of experiments on single factors may give better returns than one comprehensive experiment including many factors. Making use of the information gained from these simple trials, plans can be made for the more extensive factorial experiments which are essential to the elucidation of the interrelationships between the factors.

The discussion which follows is once again given in terms of insecticidal studies, but the applicability of the principles to other fields will be easily appreciated. Laboratory research on insecticidal potencies has been primarily directed at discovering the effect of variation in the concentration of the toxic substance on the mortality rate of the insects. To a lesser extent, the effect of variation in the duration of exposure to the poison has been examined, though published data from tests in which both factors have been varied are few. Other factors, such as the temperature or the quantity of poison used, have received even less attention.

30. QUALITATIVE FACTORS

In a well-planned toxicological investigation there will generally be at least one graded or quantitative factor, such as the concentration or amount of toxic substance, amongst the complex of factors measuring the dose. Methods of statistical analysis suitable for use when there are two or more quantitative factors are discussed in § 31 and subsequent sections. There may also be factors of a purely qualitative nature, such as variations in the medium in which the poison is applied (e.g. oil or water; Martin, 1943), or variations in the method of application of an insecticide (e.g. spray or film; Tattersfield and Potter, 1943). Other factors, though capable of being measured quantitatively, may simply be recorded descriptively, as in comparisons between 'warm' and 'cold' conditions during spraying or between 'old' and 'new' stocks of insecticide. A similar classification may be made of factors relating to the test subject.

The data from series of tests with the various combinations of conditions arising from several qualitative factors may most readily be analysed by the methods of Chapter 5 when there is only one quantitative factor, or by extensions of these when there are more. In the former case, if the probit regression lines for the quantitative factor are parallel for each combination, the effects of the qualitative factors may be measured simply by comparisons amongst their median lethal doses, and a factorial analysis of these will sort out the main effects and interaction. Yates (1937a) has discussed this type of factorial analysis, and, though some complication is introduced by the unequal precisions of the estimated log LD 50's, his work may be adapted to suit the present purpose. If the lines are not parallel, interpretation of the results is more difficult, but nothing need here be added to what has been said in § 25.

31. THE PROBIT PLANE

If the joint effects of two quantitative factors, such as the time of exposure to an insecticide and its concentration, are to be studied, batches of insects must be tested at various combinations of values of the two factors. The test conditions may be chosen as several different concentrations for each of a number of exposure times, or as several different exposure times for each of a number of concentrations. Greater symmetry may be attained by using all combinations of a set of concentrations and a set of exposure times (say 4 concentrations and 5 times used in all their 20 combinations); the more extreme combinations—low concentrations for short times, or high concentrations for long times—may be omitted if they seem unlikely to give useful results.

Experience has shown that not only is the mortality probit frequently linearly related to the logarithm of the concentration for a fixed time of exposure, but also it is frequently linearly related to the logarithm of the exposure time for a fixed concentration. As a representation of the joint effect of time and concentration, therefore, a plane may be suggested:

$$Y = \alpha + \beta_1 x_1 + \beta_2 x_2, \tag{7·1}$$

giving the probit in terms of the log concentration (x_1) and the log time (x_2). Bliss (1940b) expressed this by saying that the logarithms of the exposure time and concentration required to give any particular kill are linearly related; he proposed to evaluate this relationship by selecting a level of mortality and estimating either the values of x_1 for a series of times or the values of x_2 for a series of concentrations, according as the experiment was performed in sets of tests with fixed exposure time and varied concentration or with fixed concentration and varied time. This method of analysis involves a great amount of tedious calculation and does not give the maximum likelihood estimates of the parameters, though the estimates obtained will usually be not very different from the maximum likelihood values.

The maximum likelihood estimation is, however, to be preferred (Finney, 1943a); not only is it a more symmetrical approach and a natural generalization of the one-factor analysis, but the computations, though still lengthy, are simpler and are more easily reduced to a routine process. The method applies multiple regression analysis (Fisher, 1944, § 29) to the estimation of the coefficients β_1 and β_2, and may be further extended to three or more factors with no difficulty other than the increase in the computations. In order to use the regression technique, a provisional plane must first be fitted to the empirical probits which have been read from Table 1 in the normal manner. If several concentrations have been tested at each of a series of times, but not necessarily the same set of concentrations for every time, lines may be drawn, by eye, relating the empirical probit to the log concentration for each time, with the restriction that all the lines shall be parallel and at distances apart proportional to the differences in the log times. If the structure of the experiment is such that a series of concentrations and a series of times are used in all, or nearly all, their combinations, a method adapted from a suggestion of Richards (1941) may be used. The probits are plotted against the sum of log concentration and log time, $(x_1 + x_2)$; points with constant x_2 should then lie on one set of parallel straight lines with slope b_1, and points with constant x_1 on a second set of parallel lines with slope b_2. The differences between pairs of lines of either set will be proportional to the differences between the corresponding values of x_2 or x_1. In this way a plane representation of the three-dimensional figure relating probits to log concentration and log time is obtained, and expected probits can be read from two intersecting sets of parallel lines, drawn by eye, in the diagram. Whatever the means employed, it is worth taking some trouble over the drawing of provisional lines, in order that a satisfactory approximation to the maximum likelihood solution may be obtained by one cycle of the computations.

The working probit, y, and its weight, nw, may be derived from the expected probit and the percentage kill in precisely the same way as when only one factor is involved. Using the technique of multiple regression to derive a, b_1 and b_2 as estimates of the parameters, equation (7·1) is estimated just as was the one-factor probit equation in § 17; the details will be made clear by a careful study of Ex. 18. If this equation differs substantially from the provisional lines or plane, it may be used to determine a new series of provisional probits, with which the cycle of computations is then repeated. When there is an appreciable natural mortality amongst the controls the methods of adjustment discussed in Chapter 6 may be used; the approximate method, involving only a modification of the weighting coefficient, is directly applicable, and the full analysis described in § 28 may easily be extended to allow for an additional independent variate, x_2.

The method of estimating the parameters, testing the goodness of fit of the equation, and determining standard errors will now be illustrated by a numerical example. The statistical technique was elaborated for the analysis of data obtained by Tattersfield and Potter (1943), and Ex. 18 is a revised version of an account of the computations appropriate to the results of one of their experiments; in the account as previously published (Finney, 1942b, 1943a) weighting coefficients were not modified to allow for the control mortality.

Ex. 18.   *The effect of variation in concentration and deposit on the toxicity of a pyrethrum preparation to* Tribolium castaneum. Tattersfield and Potter (1943) have described a series of experiments on the toxicity of a solution of pyrethrum extract in heavy oil to the beetle, *T. castaneum*; the doses used consisted of all combinations of several concentrations of the pyrethrum extract and several weights of spray deposit on the glass disk on which the insects were placed. In the first experiment four concentrations (these have been measured in terms of pyrethrin I only) and three deposits were tested; the glass disk was covered with a loosely woven fabric, and each combination of concentration and deposit was used, on separate batches of insects, both as a direct spray and as a film on which the insects were afterwards placed. Batches of ten insects were used for each spraying, and all treatments were given in three-fold replication.

Insects were also exposed to different deposits of the base oil alone, applied both by the spray and by the film technique, but, as the mortalities gave no indication of being different from that amongst unsprayed controls, all control batches have been added. Of a total of 311 beetles, 12 were 'badly affected, moribund, or dead', giving a control rate of 3·9 % The full data have been reported by the experimenters (loc. cit. Table 2); after adjustment for a control death rate of 4 %, the mortalities are as shown in Table 19.

An extension of the ideas of Chapter 5 suggests the fitting of two parallel probit planes to the data. Table 20 sets out full details of the lengthy, but straightforward, computations required for this, including the derivation, from the original observations, of the figures shown in Table 19. In order to avoid

TABLE 19. Percentage Kills of *Tribolium castaneum* by a Pyrethrum Spray, adjusted for 4 % Mortality amongst the Controls (numbers of insects shown in brackets)

Pyrethrin I concentration (mg./ml.)	Deposit (mg./sq.cm.)					
	Direct spray			Film		
	0·29	0·57	1·08	0·29	0·57	1·08
0·5	0 (27)	10 (29)	17 (30)	7 (29)	11 (27)	26 (28)
1·0	50 (29)	64 (29)	61 (24)	31 (30)	48 (28)	59 (28)
2·0	90 (30)	96 (27)	100 (31)	82 (29)	96 (28)	93 (28)
4·0	100 (28)	100 (30)	100 (19)	100 (29)	100 (29)	100 (17)

the occurrence of negative numbers, both concentrations and deposits have been multiplied by 10, so that x_1 and x_2 are each 1·00 in excess of the true log concentrations and log deposits.

These modified x_1 and x_2 values form the first two columns of Table 20, and are followed by n, the number of insects under any one treatment, r, the total killed, and p', the proportionate mortality. The control mortality is estimated with considerable accuracy from the insects untreated or treated with oil alone, and, as 4 % is low and not in conflict with the remainder of the data, the approximate method of § 27 has been used without any attempt to improve the estimate of C by means of the maximum likelihood equations. The column of adjusted mortalities, p, has therefore been obtained with $C = 0·04$ in Abbott's formula (equation (6·2)), and the empirical probits of p have been written down from Table 1.

TABLE 20. Computations for Analysis of Results of Testing Various Concentrations of Pyrethrin on *Tribolium castaneum*

x_1	x_2	n	r	p'	$p(C=4)$	Empirical probit	Y	nw	y	nwx_1	nwx_2	nwy
						Exposure to direct spray						
0·70	0·47	27	1	3·7	0	$-\infty$	3·4	3·6	2·91	2·520		10·476
1·00	0·47	29	15	51·7	50	5·00	4·8	16·6	5·00	16·600		83·000
1·30	0·47	30	27	90·0	90	6·28	6·2	10·6	6·28	13·780		66·568
1·60	0·47	28	28	100·0	100	∞	7·6	1·1	7·94	1·760		8·734
								31·9		34·660	14·993	168·778
0·70	0·75	29	4	13·8	10	3·72	3·8	7·9	3·72	5·530		29·388
1·00	0·75	29	19	65·5	64	5·36	5·2	17·0	5·36	17·000		91·120
1·30	0·75	27	26	96·3	96	6·75	6·6	6·1	6·73	7·930		41·053
1·60	0·75	30	30	100·0	100	∞	8·0	0·4	8·30	0·640		3·320
								31·4		31·100	23·550	164·881
0·70	1·04	30	6	20·0	17	4·05	4·2	12·6	4·06	8·820		51·156
1·00	1·04	24	15	62·5	61	5·28	5·6	12·7	5·25	12·700		66·675
1·30	1·04	31	31	100·0	100	∞	7·0	3·9	7·42	5·070		28·938
1·60	1·04	19	19	100·0	100	∞	8·4	0·1	8·67	0·160		0·867
								29·3		26·750	30·472	147·636
								92·6		92·510	69·015	481·295
						Exposure to film						
0·70	0·47	29	3	10·3	7	3·52	3·3	3·1	3·57	2·170		11·067
1·00	0·47	30	10	33·3	31	4·50	4·7	16·7	4·51	16·700		75·317
1·30	0·47	29	24	82·8	82	5·92	6·1	11·2	5·90	14·560		66·080
1·60	0·47	29	29	100·0	100	∞	7·5	1·4	7·85	2·240		10·990
								32·4		35·670	15·228	163·454
0·70	0·75	27	4	14·8	11	3·77	3·6	5·4	3·80	3·780		20·520
1·00	0·75	28	14	50·0	48	4·95	5·0	16·5	4·95	16·500		81·675
1·30	0·75	28	27	96·4	96	6·75	6·4	8·1	6·67	10·530		54·027
1·60	0·75	29	29	100·0	100	∞	7·8	0·7	8·12	1·120		5·684
								30·7		31·930	23·025	161·906
0·70	1·04	28	8	28·6	26	4·36	4·0	9·7	4·42	6·790		42·874
1·00	1·04	28	17	60·7	59	5·23	5·4	15·8	5·22	15·800		82·476
1·30	1·04	28	26	92·9	93	6·48	6·8	4·8	6·37	6·240		30·576
1·60	1·04	17	17	100·0	100	∞	8·2	0·1	8·49	0·160		0·849
								30·4		28·990	31·616	156·775
								93·5		96·590	69·869	482·135

Spray: $\bar{x}_1 = 0.9990$, $\bar{x}_2 = 0.7453$, $\bar{y} = 5.1976$.
Film: $\bar{x}_t = 1.0330$, $\bar{x}_2 = 0.7473$, $\bar{y} = 5.1565$.

	$Snwx_1^2$	$Snwx_1x_2$	$Snwx_2^2$	$Snwx_1y$	$Snwx_2y$	$Snwy^2$
Spray	97·01900	67·43520	56·40009	502·7093	356·5278	2614·423
	92·42009	68·94792	51·43704	480·8272	358·7103	2501·565
	4·59891	$-1·51272$	4·96305	21·8821	$-2·1825$	$112·858 - 109·75 = 3·11$
Film	104·27900	70·86200	57·30655	515·5154	361·2989	2566·119
	99·78212	72·17804	52·21045	498·0687	360·2812	2486·141
	4·49688	$-1·31604$	5·09610	17·4467	1·0177	$79·978 - 75·65 = 4·33$
Total	9·09579	$-2·82876$	10·05915	39·3288	$-1·1648$	$192·836 - 183·40 = 9·44$

The empirical probits have been plotted against $(x_1 + x_2)$ in Fig. 11. Both for the direct spray and for the film technique two intersecting sets of parallel lines have been drawn, the one representing the regression of probits on x_1 for fixed x_2 and the other the regression on x_2 for fixed x_1; the lines were drawn by eye, remembering the existence of zero and 100 % kills at certain

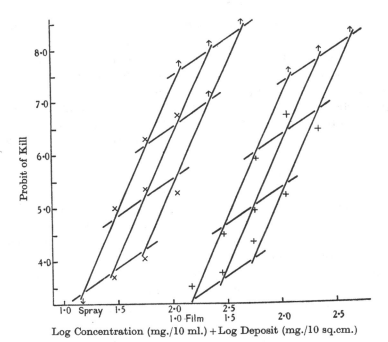

FIG. 11. Diagrammatic representation of probit planes for comparing potency of direct spray and film applications of a pyrethrum spray to *T. castaneum* (Ex. 18). × direct spray; + film. Continuous lines show effect of change in concentration at a fixed deposit. Broken lines show effect of change in deposit at a fixed concentration.

doses, in such a way as to intersect vertically above or below the plotted points. The intersections give the expected probits, Y, appropriate to each combination of x_1 and x_2, from which, by the methods of previous chapters, weights, nw, and working probits, y, were calculated; the weighting coefficients were taken from the 4 % column in Table II.

For either technique of application of the insecticide, an approximation to the maximum likelihood estimate of the probit plane is derived as the weighted regression equation of y on x_1 and x_2. In order to obtain two parallel planes, the regression coefficients must be calculated not from the sums of squares and sums of products of deviations for the techniques separately but from the totals of these (cf. § 20).

The next stage of the computations was therefore the formation of the products nwx_1, nwx_2 and nwy, the calculations proceeding independently for spray and for film; for nwx_2 it is sufficient to total the values of nw for each of the three deposits and multiply this total by the appropriate x_2, but for nwx_1 and nwy individual products must be entered for each of the twelve tests and totalled in the three deposit groups. Totals of the columns Snw, $Snwx_1$, $Snwx_2$ and $Snwy$ were then made. By summing the products of each of the twelve values of nwx_1 with x_1 and deducting $(Snwx_1)^2/Snw$, the sum of squares of deviations of x_1, denoted by $S_{x_1x_1}$, was obtained. Similarly, multiplying each of the three subtotals of nwx_1 by the corresponding x_2 adding, and deducting $(Snwx_1)(Snwx_2)/Snw$, the sum of products of deviations, $S_{x_1x_2}$, was obtained. In like manner all the sums of squares and products shown at the bottom of Table 20 for both spray and film methods have been computed, and the values for the two techniques have been added to give the last line of the table.

The regression coefficients for the two parallel planes are the solutions of the equations

$$9 \cdot 096 b_1 - 2 \cdot 829 b_2 = 39 \cdot 329,$$
$$- 2 \cdot 829 b_1 + 10 \cdot 059 b_2 = - 1 \cdot 165.$$

These should be solved by the inverse matrix method, as in Ex. 17, in order to obtain also the variances of parameters and other estimated quantities. Taking first values 1, 0 and then 0, 1 on the right-hand side, this matrix was obtained as

$$V = \begin{pmatrix} 0 \cdot 12048 & 0 \cdot 03388 \\ 0 \cdot 03388 & 0 \cdot 10894 \end{pmatrix}, \tag{7.2}$$

whence
$$\left. \begin{aligned} b_1 &= 39 \cdot 329 \times 0 \cdot 12048 - 1 \cdot 165 \times 0 \cdot 03388 \\ &= 4 \cdot 6989, \\ b_2 &= 1 \cdot 2056. \end{aligned} \right\} \tag{7.3}$$

Now the fitting of the regression planes accounts for a portion

$$b_1 S_{x_1 y} + b_2 S_{x_2 y} = 183 \cdot 40 \qquad (7 \cdot 4)$$

of the sum of squares of deviations of y, $S_{yy} = 192 \cdot 84$. S_{yy} is based on two sets of 11 degrees of freedom, and the fitted parameters remove 2, leaving $\chi^2_{[20]} = 9 \cdot 44$. This χ^2 has been analysed to provide a test of parallelism of the two planes. Regression coefficients were obtained for spray and film separately, using only the appropriate sections of the data of Table 20; the pairs of values are:

	Spray	Film
	$b_1 = 5 \cdot 1277$	$b_1 = 4 \cdot 2601$
	$b_2 = 1 \cdot 1236$	$b_2 = 1 \cdot 2999$

TABLE 21. Test of Parallelism of Probit Planes for Ex. 18

	D.F.	Sum of squares	Mean square
Parallelism of planes	2	2·00	1·00
Residual heterogeneity	18	7·44	0·41
Total	20	9·44	

The separate values of S_{yy} were then reduced by quantities calculated as in equation (7·4), namely, 109·75 and 75·65, to give residuals of 3·11 and 4·33 respectively, each being a $\chi^2_{[9]}$. Hence for a joint test of the heterogeneity of the data about two fitted planes (not now constrained to be parallel) $\chi^2_{[18]} = 7 \cdot 44$; the difference between this and the previous $\chi^2_{[20]}$ is $\chi^2_{[2]} = 2 \cdot 00$, which indicates no significant departure from parallelism. These results are summarized in Table 21. Had the heterogeneity χ^2 been large, the test of parallelism would have been based on the ratio of mean squares; the variances discussed below would then have been increased by the heterogeneity factor. Since the χ^2 values are all small, there is no need to give special consideration to possible excessive contributions from the tests at extreme dose levels, but in other circumstances the method of Ex. 7 could be employed for examining the separate contributions more carefully.

The equation to either probit plane is of the form

$$Y = \bar{y} + b_1(x_1 - \bar{x}_1) + b_2(x_2 - \bar{x}_2). \tag{7.5}$$

Substituting the appropriate values of $\bar{y}$, $\bar{x}_1$ and $\bar{x}_2$, the two fitted equations* are:

$$Y_s = -0\cdot 395 + 4\cdot 699x_1 + 1\cdot 206x_2,$$
$$Y_f = -0\cdot 598 + 4\cdot 699x_1 + 1\cdot 206x_2.$$

If the values of Y_s and Y_f are calculated for the twelve dosages used in the experiment they will be found to exhibit so good an agreement with the expected probits, Y, in Table 20 as to make it apparent that there is no necessity to repeat the cycle of computations; had a closer approximation to the solution of the maximum likelihood equations been required, Y_s and Y_f would have been used as the new expected probits.

The diagonal elements of the matrix V are the variances of b_1 and b_2, and the remaining element is the covariance between these two parameters. Hence the standard errors of b_1 and b_2 are $\pm 0\cdot 347$ and $\pm 0\cdot 330$ respectively. The kill might have been expected to be independent of the separate values of concentration and deposit, within fairly wide limits, so long as the total amount of pyrethrin (the product of concentration and deposit) remained the same. This would imply that b_1 and b_2 should only differ by an amount consistent with their sampling variation, since increases in either concentration or deposit which were equal on the logarithmic scale would have equal effects on the kill. Such equality is clearly contradicted by the results, for the variance of the difference between b_1 and b_2 is

$$V(b_1 - b_2) = 0\cdot 1205 - 2 \times 0\cdot 0339 + 0\cdot 1089$$
$$= 0\cdot 1616,$$

and therefore $\qquad b_1 - b_2 = 3\cdot 493 \pm 0\cdot 402.$

In this experiment concentration has been a far more important factor than deposit in determining the kill. In fact a doubling

---

* The equations may easily be put in terms of the logarithms of the concentrations and deposits given in Table 19 by replacing x_1 and x_2 by $(x_1 + 1)$ and $(x_2 + 1)$ respectively.

of the concentration was as effective as an increase of $(b_1 \log 2)/b_2$ in the log deposit, or a multiplication of the deposit by 14·9; the precision of this ratio is low, and 5 % fiducial limits determined by formula (4·7) from the variances and covariances in equation (7·2) are 5·9 to 279. The absorptive powers of the substratum are undoubtedly important in determining the relative effects of changes in concentration and in deposit: in further experiments with the same or a similar fabric covering the glass disk, Tattersfield and Potter again found concentration to have the greater effect, but when a hardened filter paper was used deposit became the more important factor.

In multifactorial experiments there is no unique median lethal dose; for example, with the direct spray any pair of values of x_1 and x_2 satisfying

$$4·699x_1 + 1·206x_2 = 5·395$$

is estimated to give a 50 % kill. Similarly the relative potency of the direct spray and the film technique cannot be uniquely defined, for the inequality of b_1 and b_2 implies a difference in relative potencies in respect of concentration and of deposit. The relative dosage value, or difference between equally effective dosages, may be taken as any pair of values satisfying

$$4·699M_1 + 1·206M_2 = 0·203.$$

If equal concentrations were used, log deposits 0·168 less for the spray than for the film would be expected to show equal kills, and, if equal deposits were used, equal kills would be expected when the log concentration was 0·043 less for the spray than for the film.

The mean probit difference (§ 24) was introduced (Finney, 1943a) in order to provide a single measure of the difference between two parallel probit planes. In the present experiment the value of this quantity is, by the obvious extension of equation (5·12),

$$\Delta_{sf} = Y_s - Y_f$$
$$= 0·203 \pm 0·147,$$

the variance having been obtained as

$$V(\Delta) = \frac{1}{_sSnw} + \frac{1}{_fSnw} + (\bar{x}_{1s} - \bar{x}_{1f})^2 \, V(b_1)$$

$$+ 2(\bar{x}_{1s} - \bar{x}_{1f})(\bar{x}_{2s} - \bar{x}_{2f}) \operatorname{Cov}(b_1, b_2) + (\bar{x}_{2s} - \bar{x}_{2f})^2 \, V(b_2) \quad (7 \cdot 6)$$

$$= 0 \cdot 0108 + 0 \cdot 0107 + (0 \cdot 0340)^2 \times 0 \cdot 1205$$

$$+ 2 \times 0 \cdot 0340 \times 0 \cdot 0020 \times 0 \cdot 0339 + (0 \cdot 0020)^2 \times 0 \cdot 1089$$

$$= 0 \cdot 0216.$$

32. OSTWALD'S EQUATION

If equation (7·1) be considered as expressing a relationship between values of x_1 and x_2 which give a selected kill, it may conveniently be put in the form

$$\lambda_1^{\beta_1} \lambda_2^{\beta_2} = \text{constant},$$

λ_1 and λ_2 being the absolute (not logarithmic) measures of dose, or

$$\lambda_1^{\beta_1/\beta_2} \lambda_2 = \text{constant}.$$

This equation has been used by Busvine (1938) and others, in the form

$$\lambda^n t = k, \quad (7 \cdot 7)$$

where λ is the concentration, t the time of exposure, n and k constants, as an empirical law relating the concentration and time required for a given toxic effect. Bliss (1940b) pointed out that this was a particular case of the equation

$$(\lambda - \lambda_0)^n t = k, \quad (7 \cdot 8)$$

which was used by Ostwald and Dernoschek (1910) in a discussion of the relationship between adsorption and toxic effect, λ_0 being a threshold concentration below which no effect takes place.

Bliss has only discussed experiments in which the time taken to reach 100 % kill at selected concentrations was measured. This is not usually a convenient method of studying time-concentration relationships, and is liable to give results subject to wide variation, since the time measurement is determined by the most extreme member of the batch. The more satisfactory

method is to expose batches of test subjects to concentrations
and for times chosen by the experimenter and to measure the
mortality in each batch. A threshold concentration is unlikely
to be known in advance, and would therefore have to be estimated from the data. The estimation would then become
a more troublesome process than Bliss suggests, but even
though the data require the obtaining of a probit equation
in the form

$$Y = a + b_1 \log (\lambda - \lambda_0) + b_2 \log t,$$

rather than a direct relationship between λ and t, the principle
used by him may reasonably be adopted. This entails calculating
the regression equation for several values of λ_0, but including
also a quadratic term in $\{\log (\lambda - \lambda_0)\}^2$; λ_0 is finally estimated,
by interpolation, as that value which makes the coefficient of
the quadratic term zero. This is not the maximum likelihood
estimate of λ_0 but should be a satisfactory value for practical
purposes. The maximum likelihood equations have not been
considered and are undoubtedly very complicated.

Fortunately, an adjustment to the concentration seldom
appears to be needed. If the threshold concentration regularly
differed from zero to any important extent in concentration-time
tests, the same difference would presumably occur in tests carried
out for a fixed time. Hence the adjustment would also be needed
in one-factor experiments such as have been considered in earlier
chapters, and x in equation (3·2) would have to be taken as
$\log (\lambda - \lambda_0)$ instead of $\log \lambda$, λ_0 being an additional parameter to
be estimated from the data. The fact that in so many experiments
equation (3·2) is adequate for the description of the results seems
evidence against the need of any allowance for threshold concentration. Nevertheless, some cases of data showing curvature
of the relationship between probits and log concentrations might
be simplified if a regression equation of the form

$$Y = \alpha + \beta \log (\lambda - \lambda_0) \qquad (7·9)$$

were used instead of the more usual

$$Y = \alpha + \beta \log \lambda.$$

33. The Interaction of Two Factors

The linear regression equation (7·1) implies that the effects of the two dosage factors on the mortality probit are independent and additive. If a concentration-time experiment were carried out in batches at fixed concentrations with varied times of exposure, the relationship of probit and log time might be linear for each concentration without the lines for the different concentrations necessarily being parallel. Such a situation might indicate heterogeneity of the material, or changes in the experimental conditions between the tests of different concentrations, but it might alternatively imply a true dependence of the slope of the lines upon the concentration. Bliss (1940b) has suggested that, as a first approximation, the standard deviation of the log-time tolerances (which is the reciprocal of the slope) might be expressed as a linear function of the log concentration, a method which leads to a final regression equation of Y on x_1 (log concentration) and x_2 (log time) in the form

$$Y = \frac{\alpha + \beta x_1 + x_2}{\alpha' + \beta' x_1}. \tag{7·10}$$

Equation (7·10) is linear in x_2, so that for any fixed concentration the relationship between Y and x_2 is given by a straight line, but it is not linear in x_1. It therefore differs fundamentally from the usual findings in single-factor experiments that Y is linearly related to the log concentration. The equation

$$Y = \alpha + \beta_1 x_1 + \beta_2 x_2 + \beta_{12} x_1 x_2 \tag{7·11}$$

seems preferable to (7·10), being linear in x_1 when x_2 is held constant and vice versa.* The slope of the regression line of mortality probit on either x_1 or x_2 increases (decreases if β_{12} is negative) as the other increases. The coefficient β_{12} measures the *interaction* between the two factors, or the extent to which the increase in Y for unit increase in x_1 or x_2 exceeds that predicted by the purely additive equation (7·1).

* This surface, which reduces to a plane if $\beta_{12} = 0$, is known as an hyperbolic paraboloid; sections by planes $x_1 =$ constant or $x_2 =$ constant are generators (straight lines), and sections by planes $Y =$ constant are rectangular hyperbolas.

The technique of fitting equation (7·11) to experimental data presents no difficulty. The first step is to form the products $x_1 x_2$ for each dosage level. These are then treated as the measure of a third factor of the dosage, x_3, the subsequent procedure being exactly as in Ex. 18. A multiple regression equation on x_1, x_2 and x_3 is eventually obtained in the form

$$Y = \bar{y} + b_1(x_1 - \bar{x}_1) + b_2(x_2 - \bar{x}_2) + b_{12}(x_{12} -)\overline{x_1 x_2}.$$

It should be noted that $\overline{x_1 x_2}$ is the mean value of the product $x_1 x_2$, and is different from $\bar{x}_1 \bar{x}_2$, the product of the mean values of x_1 and x_2; $\overline{x_1 x_2} = Snwx_1 x_2 / Snw$. A more complex equation such as (7·11) should not be used in preference to (7·1) unless the data clearly require it as an adequate representation of the facts, or unless there is strong *a priori* evidence of its appropriateness.

Ex. 19. *The effect of variations in concentration of hydrocyanic acid and exposure time on the mortality of* Calandra granaria. Peters and Ganter (1935) tested the toxicity of hydrocyanic acid to *C. granaria* at seven different concentrations, using batches of ten insects and two to five different exposure times for each concentration. The results, from 270 insects in all, are given in the first three columns of Table 22 and suggest that the slope of the mortality-time regression lines decreases with increasing concentration. Bliss (1940b) showed the calculations for fitting equation (7·10), but did not examine whether in fact (7·1) might not be equally satisfactory.

When equation (7·1) was fitted to the data, the result was

$$Y = -6·15 + 6·10x_1 + 6·29x_2, \tag{7·12}$$

with a residual $\chi^2_{[24]} = 53·10$; here x_1 is the log concentration in g./cu.m., and x_2 is the log exposure time in hours. Thus discrepancies from this equation show heterogeneity, with a heterogeneity factor of 2·21. Since only ten insects were used in each batch, the expected numbers of dead and surviving are necessarily small, but, as pointed out in § 18, this is less likely to disturb the χ^2 distribution than the occurrence of a few isolated cases of small expectations. Detailed examination of the separate contributions to χ^2 suggests that, though one or two are large, heterogeneity would remain after any reasonable grouping of the data. As the example is only given here in order to illustrate the technique

TABLE 22. Results of Tests of the Toxicity of Hydrocyanic Acid to *Calandra granaria*

Log concentration (g./cu.m.)	Log exposure time (hours)	% kill (p)	Empirical probit	Y
1·477	0·176	10	3·72	4·2
	0·301	40	4·75	5·0
	0·477	50	5·00	6·0
	0·544	100	∞	6·4
1·380	0·398	60	5·25	4·8
	0·477	70	5·52	5·3
	0·544	80	5·84	5·7
	0·602	90	6·28	6·1
	0·653	100	∞	6·4
1·190	0·544	50	5·00	4·3
	0·778	80	5·84	5·9
	0·903	100	∞	6·7
1·061	0·699	40	4·75	4·5
	0·778	70	5·52	5·0
	0·845	90	6·28	5·5
	1·000	100	∞	6·5
0·929	0·903	0	−∞	5·0
	0·954	40	4·75	5·4
	1·000	70	5·52	5·7
	1·079	80	5·84	6·2
	1·176	100	∞	6·9
0·778	1·204	60	5·25	6·2
	1·255	100	∞	6·6
0·544	1·204	20	4·16	4·8
	1·255	80	5·84	5·2
	1·301	90	6·28	5·6
	1·398	100	∞	6·3

At every dose $n = 10$.
The first three columns of the table are taken from Table XII of Bliss (1940b).

for fitting the regression equation, this point will not be considered further. The variance matrix

$$V = \begin{pmatrix} 0·764 & 0·693 \\ 0·693 & 0·699 \end{pmatrix} \tag{7·13}$$

must be multiplied by the heterogeneity factor to give the variances and covariance of the regression coefficients, whence

$$b_1 = 6·10 \pm 1·30, \quad b_2 = 6·29 \pm 1·24.$$

In view of the apparent decrease in the slope of separate x_2 regressions with increasing values of x_1, equation (7·11) has also been fitted to the data. Details of the computations will not be given here as they are so very similar to Ex. 18. A first set of expected probits was taken from the column headed 'Second expected probit' in Bliss's Table XII, since these values were readily available, but usually this first set would be most easily obtained from a sketch or from a simpler equation such as (7·12). Working probits and weights were found in the usual manner, and were used to calculate a weighted regression on x_1, x_2 and the product x_1x_2. The resulting equation was used to give a new set of expected probits. In all, three cycles of the computations were carried out; the expected probits for the last of these are shown as Y in Table 22. The final equation is

$$Y = -9\cdot13 + 8\cdot31x_1 + 8\cdot86x_2 - 1\cdot90x_1x_2, \qquad (7\cdot14)$$

the inverse matrix leading to the variances and covariances being

$$V = \begin{pmatrix} 2\cdot324 & 2\cdot494 & -1\cdot305 \\ 2\cdot494 & 2\cdot782 & -1\cdot511 \\ -1\cdot305 & -1\cdot511 & 1\cdot089 \end{pmatrix}. \qquad (7\cdot15)$$

The χ^2 has been calculated as in Ex. 18, by means of

$$\chi^2 = S_{yy} - b_1 S_{x_1y} - b_2 S_{x_2y} - b_{12} S_{(x_1x_2)y}.$$

This gives $\chi^2_{[23]} = 51\cdot27$, showing departures from the regression equation still to be heterogeneous, with a value of 2·23 for the heterogeneity factor. The variance of b_{12} is $2\cdot23 \times 1\cdot089 = 2\cdot428$, so that

$$b_{12} = -1\cdot90 \pm 1\cdot56.$$

The parameter b_{12} does not significantly exceed its standard error, so that no great advantage in the representation of the data arises from using equation (7·14) instead of (7·12); in other words, the data do not seriously contradict the hypothesis that $\beta_{12} = 0$.

The values of b_1 and b_2 in equation (7·14) are not directly comparable with those in (7·12). In equation (7·14) the slope of the regression on x_1 for fixed x_2 depends upon the value of x_2 and in that equation b_1 represents the slope when $x_2 = 0$; when x_2 has its mean value, the slope is

$$b_1 + b_{12}\bar{x}_2 = 6\cdot79.$$

Similarly when x_1 has its mean value, the slope of the regression on x_2 is
$$b_2 + b_{12}\bar{x}_1 = 6\cdot82.$$

These quantities are much closer to the b_1 and b_2 of equation (7·12). The variances of b_1 and b_2 obtained from the appropriate entries in the matrix (7·15) are considerably higher than the variances of b_1 and b_2 in equation (7·12), but this is of no consequence since the parameters are very different in meaning.

Since for the parameters of equation (7·12)

$$V(b_1 - b_2) = 2\cdot21 \times (0\cdot764 - 2 \times 0\cdot693 + 0\cdot699)$$
$$= 0\cdot170,$$

and therefore $b_1 - b_2 = -0\cdot19 \pm 0\cdot41,$

there is no significant difference between the two regression coefficients, and a regression equation with $(x_1 + x_2)$ as the single independent variate would represent the data almost as satisfactorily as equation (7·14). Hence, in this experiment, the toxic effect of the hydrocyanic acid may be expressed purely in terms of the number of 'grams-per-cubic-metre hours', the product of the concentration and the time of exposure. Taking the sum of x_1 and x_2 as a new variate, x, the methods of § 17 would lead to the regression equation, which may be derived with sufficient accuracy from sums of squares and products already computed as

$$Y = -6\cdot32 + 6\cdot27(x_1 + x_2),$$

with $\chi^2_{[25]} = 53\cdot57$

and $b = 6\cdot27 \pm 1\cdot22.$

34. EXTENSIONS TO SEVERAL FACTORS

The methods outlined in this chapter may easily be applied to data relating to more than two dosage factors, though, as always in multiple regression analysis, the amount of computing involved increases rapidly with increasing number of factors. For three factors, equation (7·1) may be extended to the form

$$Y = \alpha + \beta_1 x_1 + \beta_2 x_2 + \beta_3 x_3, \qquad (7·16)$$

and, though expected probits are less easily obtained, the estimation of the parameters follows the same procedure as before. On the analogy of equation (7·11), a more general equation, linear in each factor separately but allowing for interactions between them, is

$$Y = \alpha + \beta_1 x_1 + \beta_2 x_2 + \beta_3 x_3 + \beta_{12} x_1 x_2$$
$$+ \beta_{13} x_1 x_3 + \beta_{23} x_2 x_3 + \beta_{123} x_1 x_2 x_3, \qquad (7·17)$$

in which the regression coefficient β_{12} measures the interaction between the first two factors, and β_{123} measures the three-factor interaction. To fit such an equation to experimental results, the four products $x_1 x_2$, $x_1 x_3$, $x_2 x_3$ and $x_1 x_2 x_3$ are treated as though they measured separate factors, and a seven-variate regression equation is calculated.

As in the unifactorial analysis, the cycle of computations for fitting a regression equation is repeated until satisfactory agreement with the maximum likelihood estimates is obtained, as shown by the agreement between successive sets of expected probits. Careful choice of the first set will often ensure that one or, at most, two cycles suffice. Tests of heterogeneity and of parallelism are easily made, and standard errors of parameters are derived from the inverse matrix used in solving the equations for the estimates. The notion of relative potency, so useful in tests of one factor, has no simple multifactorial analogue, but in all cases the mean probit difference can be used for comparing series of parallel results; 'parallel' is here used with the meaning that equations such as (7·16) or (7·17) can adequately describe the data, with the restriction that corresponding regression coefficients, β, shall be the same for every series and thus that the equations for the several series shall differ only in their values of α.

Chapter 8

THE TOXIC ACTION OF MIXTURES OF POISONS

35. TYPES OF JOINT ACTION

ANY attempt to understand fully the toxic action of a group of insecticides or fungicides must ultimately involve a study of their behaviour when two or more are applied in mixture. In some cases the potency of a mixture may be greater than would be expected simply from a knowledge of the potencies of the constituents separately, a result which is clearly of practical importance in the economic utilization of the poisons; the opposite situation of reduced potency of a mixture by comparison with that of its constituents may also occur. Precise meaning can only be given to these modes of action of mixtures after the establishment or definition of a normal mode of action; the results of any series of tests may then be compared with this standard in order to judge whether the toxicity is enhanced or reduced when the poisons are applied in mixture.

The first systematic discussion of this topic in relation to probit analysis was given by Bliss (1939a) in a paper on 'The toxicity of poisons applied jointly'. He distinguished three types of joint toxic action, independent, similar and synergistic, whose properties he described as:

'(1) *Independent joint action.* The poisons or drugs act independently and have different modes of toxic action. The susceptibility to one component may or may not be correlated with the susceptibility to the other. The toxicity of the mixture can be predicted from the dosage-mortality curve for each constituent alone and the correlation in susceptibility to the two poisons; the observed toxicity can be computed on this basis whatever the relative proportions of the components.

'(2) *Similar joint action.* The poisons or drugs produce similar but independent* effects, so that one component can be

---

* An unfortunate choice of word, since the meaning is entirely different from that in the previous paragraph.

substituted at a constant proportion for the other; variations in individual susceptibility to the two components are completely correlated or parallel. The toxicity of a mixture is predictable directly from that of the constituents if their relative proportions are known.

'(3) *Synergistic action.* The effectiveness of the mixture cannot be assessed from that of the individual ingredients but depends upon a knowledge of their combined toxicity when used in different proportions. One component synergizes or antagonizes the other.'

Bliss explicitly excluded from this classification cases of two constituents which react chemically to form a new compound. In one sense he may be considered to have covered all other types of joint action, independent and similar action being the simplest, and synergism including all forms of departure from the normal. *Antagonistic action,* in which the potency of a mixture is less than expected, has been described by Clark (1937, Chapter 17), who suggested various mathematical representations of it; his examples, however, are of so different a nature as scarcely to fall within the scope of Bliss's paper or the present.work. Antagonism will be treated here simply as negative synergism.

The potency of a mixture whose constituents act similarly is generally greater than that of a mixture, in the same proportions, whose constituents are of the same individual potencies but act independently (§ 38). Either type of action is specified by an exact law predicting the kill produced by a mixture from the amounts of the constituents and their potencies. Hence a more exact definition of synergism is needed than Bliss's statement of its being 'characterized by a toxicity greater than that predicted from experiments with the isolated constituents'; at least it must be decided whether independent, or similar, or some other joint-action law is the norm to which any suspected case of synergism is to be referred. Bliss suggested two alternative mathematical models for synergistic action, but neither of these nor any of those used by Clark has the more familiar concepts of independent or similar action as special cases of zero synergism.

Finney (1942a) has endeavoured to bring out the logical relationship between different types of joint action. That of chief importance in studies of insecticides and fungicides appears to be similar action, which will therefore be first considered here.

36. SIMILAR ACTION

As has been said in § 20, two poisons whose modes of action on the test organism are much alike, especially poisons of related chemical constitutions, often show parallel regression lines of mortality probits on log doses. The relative potency can then be expressed by a single figure, the ratio of equally effective doses, which is a constant at all levels of mortality. If the two regression lines are written as

$$Y_1 = a_1 + b \log \lambda, \tag{8.1}$$

$$Y_2 = a_2 + b \log \lambda, \tag{8.2}$$

where λ is the dose, the potency of the second poison relative to that of the first is given by

$$\log \rho_2 = (a_2 - a_1)/b, \tag{8.3}$$

so that (8.2) may alternatively be written

$$Y_1 = a_1 + b \log (\rho_2 \lambda).$$

Multiplication by the factor ρ_2 converts doses of the second poison into equivalent doses of the first, the kill then being predictable by means of equation (8.1). A mixture containing amounts λ_1, λ_2 of the two poisons is said to show similar action if, within the limits of sampling variation, the kill is the same as that which would be produced by a dose of the first equal to the sum of λ_1 and $\rho_2 \lambda_2$; thus similar action requires that the probit regression line for a mixture shall have the form

$$Y = a_1 + b \log (\lambda_1 + \rho_2 \lambda_2). \tag{8.4}$$

If the mixture is applied as a total dose, λ, in which the proportions of the two poisons are π_1, π_2, equation (8.4) may conveniently be rewritten as

$$Y = a_1 + b \log (\pi_1 + \rho_2 \pi_2) \lambda.$$

Though it is not necessary that $\pi_1 + \pi_2 = 1$, since λ may be measured in terms of some preparation containing also a proportion of inactive materials, equations (8·1) and (8·2) can be obtained by taking $\pi_1 = 1$, $\pi_2 = 0$ and vice versa. The potency of the mixture relative to that of the first poison is

$$\rho = \pi_1 + \rho_2 \pi_2, \qquad (8·5)$$

and $(\pi_1 + \rho_2 \pi_2)\lambda = \lambda_1 + \rho_2 \lambda_2$ is the expression of a dose, λ, of the mixture as an equivalent, or equally effective, dose of the first poison. From these relationships it follows that if Λ_1, Λ_2 are the ED 50's for the two poisons, $\Lambda_2 = \Lambda_1/\rho_2$ and, more generally, for any mixture the ED 50 is estimated as

$$\Lambda = \Lambda_1/(\pi_1 + \rho_2 \pi_2). \qquad (8·6)$$

The equations (8·4) and (8·6) are the same as (5) and (7) of Bliss's paper, except for the changed notation.

The concept of similar action and of equivalent doses is easily extended to three or more poisons, provided that all may be taken to have the same probit-dosage regression coefficient, b. If a mixture contains quantities λ_1, λ_2, λ_3 of three such poisons, the second and third having potencies ρ_2, ρ_3 relative to the first, the equivalent dose is $(\lambda_1 + \rho_2 \lambda_2 + \rho_3 \lambda_3)$ and the regression line for the mixture

$$Y = a_1 + b \log (\lambda_1 + \rho_2 \lambda_2 + \rho_3 \lambda_3). \qquad (8·7)$$

If the regression equations for two poisons which exhibit similar action and for a mixture of the two in proportions $\pi_1 : \pi_2$ $(\pi_1 + \pi_2 = 1)$ are written as

$$Y_1 = a_1 + bx, \quad Y_2 = a_2 + bx, \quad Y_3 = a_3 + bx, \qquad (8·8)$$

the potency of the second poison relative to the first is estimated as $\rho = 10^M$, where
$$M = (a_2 - a_1)/b;$$

the third regression equation must then be equivalent to

$$Y_3' = a_1 + b \log (\pi_1 + \rho \pi_2) + bx, \qquad (8·9)$$

apart from sampling errors in the estimation of the four parameters a_1, a_2, a_3, b.

Ex. 20. *Similar joint action of rotenone and a deguelin concentrate.* Data from tests of rotenone, a deguelin concentrate, and a mixture of the two in the proportion of 1 : 4 have been examined in Ex. 10, and have been found to be satisfactorily fitted by the three parallel probit lines shown as equations (5·8):
$$Y_r = 2{\cdot}336 + 3{\cdot}891x, \quad Y_d = 0{\cdot}660 + 3{\cdot}891x, \quad Y_m = 1{\cdot}302 + 3{\cdot}891x.$$
From the first two of these, the potency of the deguelin concentrate relative to that of rotenone is assessed, as in Ex. 10, at
$$\rho_d = 10^{-0{\cdot}4307} = 0{\cdot}371.$$
If the two poisons act similarly, a dose λ of the mixture will be as effective as a dose
$$(\pi_r + \rho\pi_d)\,\lambda = (0{\cdot}2 + 0{\cdot}371 \times 0{\cdot}8)\,\lambda = 0{\cdot}497\lambda$$
of rotenone, from which the regression line for the mixture may be predicted as
$$Y'_m = 2{\cdot}336 + 3{\cdot}891 \log 0{\cdot}497 + 3{\cdot}891x$$
$$= 1{\cdot}155 + 3{\cdot}891x.$$

Comparison of this prediction with the equation estimated directly from the data shows the toxicity of the mixture to be a little greater than that required by the hypothesis of similar action. The logarithm of the observed potency relative to that predicted is
$$M_s = (1{\cdot}302 - 1{\cdot}155)/3{\cdot}891$$
$$= 0{\cdot}0378;$$

thus the mixture is estimated to be 9 % more toxic than if the constituent poisons acted similarly. This apparent slight synergistic effect is shown in Ex. 21 to be within the limits of sampling variation, and no significance need be attached to it.

The difference between the third equation of (8·8) and equation (8·9), the probit-dosage relationship estimated directly from the data and the prediction according to the hypothesis of similar action respectively, is a mean probit difference measuring the amount of any enhancement or reduction of effectiveness, or, in general terms, a measure of synergism; it is
$$\Delta_s = a_3 - a_1 - b \log{(\pi_1 + \rho\pi_2)}. \tag{8·10}$$

The variance of Δ_s has a complicated expression, but is not difficult to calculate from quantities already used in estimating the various parameters:

$$V(\Delta_s) = \frac{1}{(\pi_1+\rho\pi_2)^2}\left[\frac{\pi_1^2}{_1Snw}+\frac{\rho^2\pi_2^2}{_2Snw}+\frac{(\pi_1+\rho\pi_2)^2}{_3Snw}\right.$$
$$\left.+\frac{\{\pi_1\bar{y}_1+\rho\pi_2\bar{y}_2-(\pi_1+\rho\pi_2)(\bar{y}_3-\Delta_s)\}^2}{b^2\Sigma S_{xx}}\right], \quad (8\cdot11)$$

suffices indicating the data for the three preparations tested, and Σ indicating summation for the three. As always, the variance must be multiplied by the heterogeneity factor when this is significantly greater than unity. Comparison of Δ_s with its standard error, the square root of the expression in (8·11), gives a test of the significance of the departure from the similar action prediction; positive values of Δ_s indicate synergism, negative antagonism.

The quantity $\quad M_s = \dfrac{a_3-a_1}{b} - \log(\pi_1+\rho\pi_2) \quad (8\cdot12)$

is the logarithm of the ratio of the observed potency of the mixture to that predicted on the hypothesis of similar action. Since
$$M_s = \Delta_s/b,$$

the sign of M_s also shows whether the departure from similar action is in the direction of synergism or of antagonism, but the test of significance should be made on Δ_s rather than on M_s. Should the standard error of M_s also be required, it may be obtained from the variance

$$V(M_s) = \frac{1}{b^2(\pi_1+\rho\pi_2)^2}\left[\frac{\pi_1^2}{_1Snw}+\frac{\rho^2\pi_2^2}{_2Snw}+\frac{(\pi_1+\rho\pi_2)^2}{_3Snw}\right.$$
$$\left.+\frac{\{\pi_1\bar{y}_1+\rho\pi_2\bar{y}_2-(\pi_1+\rho\pi_2)\bar{y}_3\}^2}{b^2\Sigma S_{xx}}\right]. \quad (8\cdot13)$$

Exact formulae for the fiducial limits of M_s and Δ_s have not been developed, but, unless g (as defined in §19) is large, limits calculated from the standard error should be sufficiently reliable for Δ_s and probably also for M_s. The error committed in assessing the fiducial limits of M_s as though the estimate were normally distributed will usually be less serious than in similar estimations of fiducial limits for median lethal dosages or relative dosage values; on account of the greater number of tests, the accuracy of estimation of b will be increased and g will therefore be reduced.

Ex. 21. *Standard errors of discrepancies between observation and similar action predictions*. The mean probit difference between the regression equation for the rotenone and deguelin concentrate mixture discussed in Ex. 20 and the equation predicted on the hypothesis of similar action is

$$\Delta_s = 1 \cdot 302 - 1 \cdot 155 = 0 \cdot 147.$$

Also, using numerical values found in Ex. 10,

$$\frac{\pi_r \bar{y}_r + \rho \pi_d \bar{y}_d}{\pi_r + \rho \pi_d} - \bar{y}_m = -0 \cdot 403,$$

and therefore, from equation (8·11), remembering that no evidence of heterogeneity was found when these data were analysed in Ex. 10,

$$V(\Delta_s) = \frac{(0 \cdot 2)^2}{119 \cdot 6 \times (0 \cdot 497)^2} + \frac{(0 \cdot 297)^2}{78 \cdot 1 \times (0 \cdot 497)^2} + \frac{1}{74 \cdot 4} + \frac{(0 \cdot 403 - 0 \cdot 147)^2}{15 \cdot 138 \times 10 \cdot 467}$$

$$= \frac{0 \cdot 0014638}{0 \cdot 24701} + 0 \cdot 01344 + \frac{0 \cdot 065536}{158 \cdot 45}$$

$$= 0 \cdot 01978.$$

Hence
$$\Delta_s = 0 \cdot 147 \pm 0 \cdot 141,$$

a value which is not significantly different from zero and which therefore shows that the indication of synergism is within the limits of sampling variation.

The calculations for $V(\Delta_s)$ have been shown in some detail, as by arranging them in this way some steps can be used again in calculating the variance of the log ratio of potencies, M_s.

From equation (8·13)

$$V(M_s) = \left[\frac{0 \cdot 0014638}{0 \cdot 24701} + 0 \cdot 01344 + \frac{(0 \cdot 403)^2}{158 \cdot 45} \right] \Big/ 15 \cdot 138$$

$$= 0 \cdot 001347.$$

The value of M_s has already been found in Ex. 20, whence

$$M_s = 0 \cdot 0378 \pm 0 \cdot 0367.$$

Fiducial limits to M_s are approximately, for 5 % probability, $0 \cdot 0378 \pm 1 \cdot 96 \times 0 \cdot 0367$ or $-0 \cdot 0341$ and $0 \cdot 1097$, so that the true potency of the mixture may be expected to lie between 92·4 and 128·7 % of the similar action prediction.

The test of significance of $\varDelta_s$ may be carried out as a χ^2 test, taking
$$\chi^2_{[1]} = \varDelta_s^2 / V(\varDelta_s), \tag{8·14}$$

this test being precisely equivalent to the test based on the normal distribution and the standard error of $\varDelta_s$. If tests have been carried out on several mixtures of two poisons of different proportionate constitution, each giving an estimated $\varDelta_s$, a composite test of agreement with the similar action hypothesis may be obtained by adding the χ^2 values calculated as in (8·14) and testing as a χ^2 with the total number of degrees of freedom. If allowance has had to be made for heterogeneity, the total of the χ^2 values must be divided by the degrees of freedom and the result used in a variance ratio test of significance. The various $\varDelta_s$ and $V(\varDelta_s)$ are, strictly speaking, not independent, since all are dependent upon the common value of b and upon information from tests on the constituent poisons used separately. Nevertheless, the composite test derived from equation (8·14) should give an approximate test of synergism for use when the more complicated procedure described in § 37 seems unnecessary.

In any experiment designed to throw light on the potency of mixtures of poisons, the constituent poisons should be tested alone as well as in mixture. Even when a considerable amount of information on their toxic effects is already available, they should not be omitted from further tests. The insects or other test subjects used in work of this nature often show great variability in response, both amongst themselves and as a result of uncontrolled experimental conditions; hence, unless it is certain that conditions have not changed appreciably in any important aspect, tests of mixtures cannot safely be compared with tests of the separate constituents made on a different occasion.

The internal consistency of results for three or more mixtures of two poisons with the hypothesis of similar action may, however, be judged, even without tests on the poisons separately, by expressing intermediate members of the series as though they were obtained by mixing, in the appropriate proportions, the two mixtures of most extreme constitution; the illustration given in Ex. 22 will suffice to show how this is done.

Ex. 22. *Expression of one mixture in terms of two others.* Suppose tests have been made on three mixtures, containing respectively proportions 4 : 1, 7 : 3, 2 : 3 of two poisons. The first and the last are the most extreme in composition, and the second may easily be expressed in terms of them. The proportions of the two constituent poisons in the mixtures are $(0.8, 0.2)$, $(0.7, 0.3)$ and $(0.4, 0.6)$; the second mixture may therefore be considered as composed of proportions π_1, π_2 of the first and third where

$$0.8\pi_1 + 0.4\pi_2 = 0.7, \quad 0.2\pi_1 + 0.6\pi_2 = 0.3,$$

whence $\pi_1 = 0.75$, $\pi_2 = 0.25$. It is easily verified that if the 4 : 1 and 2 : 3 mixtures are themselves mixed in the ratio 3 : 1 a mixture of the original poisons in the ratio 7 : 3 is obtained. The agreement of the results of the experiment with predictions made from the hypothesis of similar action may then be examined by means of the methods and formulae just discussed.

37.   A GENERAL TEST FOR SIMILAR ACTION

When several mixtures with different proportionate constitutions have been tested, another method (Finney, 1942a) may be used for examining the agreement of the median effective doses with the values predicted by similar action. This method explicitly uses only the median effective doses and their standard errors; parallelism of the regressions is implied, however, since otherwise no meaning can be given to the concept of similar action. There are theoretical objections to the method on account of certain assumptions of normality and independence, but providing that the regression coefficient is estimated with reasonable precision the resulting tests of significance should be reliable.

From equation (8·6), if λ_1 is the LD 50 of a poison, ρ the relative potency of a second poison, and π_1, π_2 the proportions of the two poisons in a mixture, the LD 50 of the mixture is λ, where

$$\frac{1}{\lambda} = \left(\frac{1}{\lambda_1}\right)\pi_1 + \left(\frac{\rho}{\lambda_1}\right)\pi_2. \qquad (8\cdot15)$$

Consequently, if values of λ have been estimated for mixtures of two poisons in a series of different proportions, $1/\lambda_1$ and ρ/λ_1 may be estimated as the regression coefficients of $1/\lambda$ on π_1 and π_2, each value of $1/\lambda$ being weighted inversely as its variance. If λ is the LD 50 corresponding to a log dose m ($m = \log_{10}\lambda$)

$$V(1/\lambda) = \frac{V(m)}{\lambda^2}(\log_e 10)^2 = \frac{5\cdot302 V(m)}{\lambda^2},$$

and therefore the weight to be attached to $1/\lambda$ is

$$W = 0\cdot1886\lambda^2/V(m). \qquad (8\cdot16)$$

The regression equation (8·15) contains no 'constant term', but is constrained to give λ infinite when the content of both constituents is zero; hence total sums of squares and products must be used in calculating the regression coefficients, not sums adjusted so as to refer to deviations about means. Comparison of the values of λ calculated for each mixture from equation (8·15) with the values from which the equation was estimated shows how well the observations are fitted by the similar action hypothesis and permits a test of significance of the agreement.

Tattersfield and Martin (1935; see also Ex. 13 above) have published results of toxicity tests with ether extracts of seven different derris roots to *Aphis rumicis*, and have given the LD 50 for each in terms of rotenone and a dehydro mixture. The latter is known (Martin, private communication) to have varied considerably from root to root in its relative proportions of different dehydro compounds. Though he recognized that any comparison of toxicities based on the two constituents alone must be of doubtful value, Bliss used these data as an illustration of synergistic action between two poisons and found them to agree satisfactorily with a generalization of one of his formulae for synergism, provided that one root (no. 2), which behaved anomalously, was omitted. On account of the non-independence of the LD 50's and their standard errors, the analysis given is not strictly correct, but the disturbance so caused can only be slight. A more serious error of tactics is that the data were not examined for their possible agreement with the simpler hypothesis of similar action. Finney (1942a) has shown the median lethal doses to be satisfactorily fitted by equation (8·15), using the regression procedure outlined above. Details of the calculations will not be shown here, since the process is sufficiently illustrated by Ex. 23, in which the method is used to test similar action between three components.

Ex. 23. *Similar action between the toxic constituents of derris root.* Results of toxicity tests on four derris roots, carried out by Martin (1940), have been examined in Ex. 14, and comparative LD 50 values have been estimated, allowing for different levels of susceptibility of the test insects on the three days of testing; these values are given in Table 15. The variances of the median lethal doses were not discussed in Ex. 14, but for present purposes the sum of the weights in the appropriate column of Table 14 will be taken as the weight to be attached to any log LD 50; a little consideration shows this figure to be too high, though it should be of the right order of magnitude. Allowance must also be made for the heterogeneity of the relative potencies which was demonstrated by a χ^2 test at the end of Ex. 14, but this can most conveniently be done at a later stage of the analysis.

Martin subdivided the toxic constituents of each of the four roots into rotenone, a toxicarol fraction (possibly including sumatrol, malaccol, and other materials in addition to toxicarol) and a deguelin concentrate fraction (including elliptone, if present, as well as deguelin), and determined the proportions of these three for each root. The average probit regression coefficients for the three days of testing were $5·77 \pm 0·39$, $5·07 \pm 0·35$ and $4·66 \pm 0·32$. Though differences between these are not significant, there is some indication of a decrease from the beginning to the end. There is no evidence of consistent differences between roots in their log-tolerance variances. An examination of the adequacy of similar action to explain the observations is therefore of interest and provides a useful illustration of the method. If the three components act similarly, equation (8·7) should be adequate to describe the observed mortalities, and the LD 50 for any root should be given by

$$\frac{1}{\lambda} = \left(\frac{1}{\lambda_r}\right)\pi_r + \left(\frac{\rho_t}{\lambda_r}\right)\pi_t + \left(\frac{\rho_d}{\lambda_r}\right)\pi_d, \tag{8·17}$$

π_r, π_t, π_d being the proportions of the three components in the root, λ_r the LD 50 of rotenone, and ρ_t, ρ_d the potencies of the toxicarol and deguelin fractions relative to rotenone.

TABLE 23.   Computations for Test of Similar Action between the Constituents of Derris Root

Material	$1/V(m)$	λ	W	π_r	π_t	π_d
Rotenone	1,640	13·2	$0·054 \times 10^6$	1·0000	0·000	0·000
W. 211	6,710	244	$75·3 \times 10^6$	0·0146	0·152	0·086
W. 212	6,140	146	$24·7 \times 10^6$	0·0414	0·043	0·124
W. 213	4,710	188	$31·4 \times 10^6$	0·0346	0·024	0·082
W. 214	4,820	92·0	$7·69 \times 10^6$	0·0794	0·026	0·122

Material	$1/\lambda$	$W\pi_r$	$W\pi_t$	$W\pi_d$	W/λ
Rotenone	0·07576	54,000	0	0	4,091·04
W. 211	0·00410	1,099,380	11,445,600	6,475,800	308,730
W. 212	0·00685	1,022,580	1,062,100	3,062,800	169,195
W. 213	0·00532	1,086,440	753,600	2,574,800	167,048
W. 214	0·01087	610,586	199,940	938,180	83,590·3

Table 23 shows the first stage in the fitting of this equation to the data for rotenone alone and for the four roots. The column $1/V(m)$ contains the column totals from Table 14, representing approximately the weight to be attached to each log LD 50, and the column λ contains the values of the LD 50 from Table 15; from these and equation (8·16), W has been calculated. The proportions of rotenone, toxicarol and deguelin (π_r, π_t and π_d) are reproduced from Martin's Table 12 (1940), $1/\lambda$ is tabulated, and the products of W with each of the four following columns are then entered in full.

Sums of products of the entries in each of the last four columns in Table 23 with π_r, π_t, π_d and $1/\lambda$ are next formed. Sums of products such as $SW\pi_r\pi_t$ are obtained in two ways, the agreement being a check on the arithmetic; sums of squares such as $SW\pi_r^2$ must be recalculated as a check. Systematic arrangement of the results gives the following equations for the three regression coefficients in (8·17):

$$198{,}457(1/\lambda_r) + 253{,}026(\rho_t/\lambda_r) + 384{,}926(\rho_d/\lambda_r) = 28{,}020{\cdot}1,$$

$$253{,}026(1/\lambda_r) + 1{,}808{,}686(\rho_t/\lambda_r) + 1{,}202{,}210(\rho_d/\lambda_r) = 60{,}384{\cdot}8,$$

$$384{,}926(1/\lambda_r) + 1{,}202{,}210(\rho_t/\lambda_r) + 1{,}262{,}298(\rho_d/\lambda_r) = 71{,}426{\cdot}9.$$

These equations must be solved to a greater number of decimal places than is needed for the comparison of potencies, in order to have sufficient accuracy in the test of significance of the departure from similar action. The solutions are:

$$1/\lambda_r = 0{\cdot}0800010, \quad \rho_t/\lambda_r = 0{\cdot}0021760, \quad \rho_d/\lambda_r = 0{\cdot}0301168.$$

Also the weighted sum of squares of $1/\lambda$ is

$$SW(1/\lambda)^2 = 4532{\cdot}04,$$

and the portion of this accounted for by the linear regression on π_r, π_t, π_d is

$$(1/\lambda_r)\,SW\pi_r/\lambda + (\rho_t/\lambda_r)\,SW\pi_t/\lambda + (\rho_d/\lambda_r)\,SW\pi_d/\lambda = 4524{\cdot}18.$$

Now $SW(1/\lambda)^2$, being a sum of squares *not* adjusted for a mean, has 5 degrees of freedom, and the residual after the fitting of three constants therefore has 2 degrees of freedom. If the weights, W, had been truly the reciprocals of the variances of $1/\lambda$, the residual would have been a χ^2, namely,

$$\chi^2_{[2]} = 7{\cdot}86.$$

Quite apart from the considerations of non-independence mentioned earlier, the weights require to be reduced on account of the heterogeneity of relative potencies found in Ex. 14

$$\chi^2_{[2]} = 14{\cdot}24.$$

Hence a test of significance for departures from the similar action law is obtained by comparing mean squares from these two χ^2 values (Fisher and Yates, 1943, Table V). Since the mean square for departures from similar action is less than the heterogeneity mean square, the data do not contradict similar action, but the test is not very precise because of the few degrees of freedom available for the heterogeneity χ^2 and mean square.

The LD 50 for rotenone, λ_r, is estimated as 12·50 mg./l., and the potencies of the toxicarol and deguelin fractions relative to rotenone, ρ_t and ρ_d, are estimated as 0·02720 and 0·37645 or $\frac{1}{37}$ and a little over $\frac{1}{3}$, as compared with the values of $\frac{1}{15}$ and $\frac{1}{5}$ used by Martin in computing rotenone equivalents for the four roots. The rotenone equivalent is $(\pi_r + \rho_t \pi_t + \rho_d \pi_d)$, and the estimated LD 50 for any root is obtained by dividing the LD 50 for

TABLE 24.   Comparison of Median Lethal Doses of Derris Roots and Values Predicted according to Similar Action (doses in mg./l.)

Sample	LD 50 in Table 15	Predicted from all data		Predicted, omitting rotenone	
		% rotenone equivalent	LD 50	% rotenone equivalent	LD 50
Rotenone	13·2	100·000	12·5	100·000	(9·09)
W. 211	244	5·111	245	3·728	244
W. 212	146	8·925	140	6·240	146
W. 213	188	6·612	189	4·819	189
W. 214	92·0	12·603	99·2	9·898	91·9

rotenone by the rotenone equivalent. Table 24 shows the results and makes it clear that the expression of the toxic contents of the four roots in terms of their rotenone equivalents gives a very reasonable evaluation of their potencies.

The rotenone content of the roots is so small, and in consequence the potencies of the roots are so very different from that of rotenone, that it may seem more appropriate to examine the consistency with the similar action law of the data for the four roots alone, ignoring the rotenone tests entirely. The only changes which the omission of the first line in Table 23 produces in the equations for the three parameters are that $SW\pi_r^2$ becomes 144,457 instead of 198,457 and $SW\pi_r/\lambda$ becomes 23,929·1 instead of 28,020·1. The solutions now lead to

$$\lambda_r = 9·093, \quad \rho_t = 0·06642, \quad \rho_d = 0·14634.$$

That the four roots show excellent agreement with the similar action hypothesis is evident from the rotenone equivalents and estimated LD 50's, also shown in Table 24, though naturally the discrepancy for rotenone itself becomes much greater than in the earlier analysis.

From his examination of the same data, Martin concluded that there was some evidence of synergistic action between the three toxic constituents of the roots. The present more complete analysis shows that the indications of synergism disappear when the best possible values for the relative potencies of the constituents are estimated from the data themselves instead of from an earlier investigation. Nevertheless, synergism may be present in a form that leaves the data internally consistent with similarity; the matter can scarcely be settled satisfactorily without comprehensive trials including tests both on the roots and on the constituents.

38. INDEPENDENT ACTION

The distinction between similar and independent action must be kept clear. In mixtures whose constituents act similarly any quantity of one constituent can be replaced by a proportionate amount of any other without disturbing the potency, but for mixtures whose constituents act independently the mortalities, not the doses, are additive. This type of action may occur with a mixture whose constituents produce their toxic effect in entirely different ways, as, for example, a mixture of two insecticides of which one is a stomach poison and the other a contact poison.

Suppose that the doses of two poisons given in a mixture are capable of producing mortalities P_1, P_2 when used separately. If the two act independently, a proportion P_2 of the test subjects which would survive the first poison is expected to succumb to the second, thus giving an expected total mortality (cf. § 26)

$$P = P_1 + P_2(1 - P_1).$$

This may be written

$$P = 1 - (1 - P_1)(1 - P_2) \tag{8.18}$$

in order to display the symmetry in P_1 and P_2. Equation (8.18) is the basic expression of independent action between two poisons; for three constituents, the corresponding relationship is

$$P = 1 - (1 - P_1)(1 - P_2)(1 - P_3), \qquad (8 \cdot 19)$$

whence the method of extension to a greater number of constituents is apparent.

Bliss (1939a) has suggested a modification of equation (8·18) to allow for a correlation in the susceptibility to the two components. If $P_1 > P_2$, so that the first component is the more toxic, his equation is

$$P = P_1 + P_2(1 - P_1)(1 - r), \qquad (8 \cdot 20)$$

r representing the degree of correlation between the susceptibilities to the two poisons.* When $r = 0$, this last equation reduces to (8·18); when $r = 1$ it becomes simply

$$P = P_1,$$

so that the combined mortality is the same as that for the more toxic ingredient applied alone. Negative correlation of susceptibilities does not fall within the scope of equation (8·20), since negative values of r would make $P > 1$ for certain values of P_1 and P_2. As will be seen from Ex. 26, a critical experiment to distinguish between equations (8·18) and (8·20), even for values of r close to 1, would usually require many more test subjects than are normally available; there appears to be no evidence, either theoretical or experimental, that equation (8·20) represents any real biological situation, and therefore no detailed study of it will be attempted here. Indeed, so far as is known to the writer, no clear experimental demonstration of the occurrence of the simplest form of independent action has yet been made, and equation (8·18) may well be too crude an approximation to any interaction of effects of mixed poisons to be of much practical value. A discussion of the types of relationship to which the equation can lead is of interest, however, as giving some indication of what may be expected in a future approach to more complex problems.

Simple though the concept of independent action is, the statistical treatment of data relating to poisons acting in this

* The symbol r is used here with a meaning entirely different from that in other sections of this book.

manner is much more difficult than that of data relating to similar action. Even though the two constituents of a mixture give mortality probits which are linearly related to the log dose and the two probit regression lines are parallel, the regression relationship for the mixture will not be a straight line; still less can this be so when the regression lines for the constituents are not parallel. No exact methods of statistical analysis are available, but a few illustrations of curves obtained according to the independent action law, for mixtures whose constituents have the usual normal distribution of log tolerances, will show the types of probit-dosage relationship which may be encountered.

Ex. 24.   *Independent action between constituents with parallel probit regression lines.* Consider two poisons having the probit regression lines
$$Y_1 = 6 + 2x, \quad Y_2 = 4 + 2x;$$

the first has ten times the potency of the second at all levels of mortality. In a $1:1$ mixture of the two, the concentration of the first poison in a total concentration λ will be $\frac{1}{2}\lambda$, and the log concentration of the first will therefore be $(x - \log 2)$. For low values of x, the kill produced by the second poison will be negligible relative to that produced by the first, and the total effect will be almost identical with that caused by the content of the first poison alone; hence for low values of x the mixture gives mortality probits determined by
$$Y = 6 + 2(x - \log 2)$$
$$= 5\cdot40 + 2x.$$

With increasing total concentration, the amount of the second poison present in the mixture begins to produce an appreciable effect, so that the total kill is greater than that for the first alone. The curve relating the probit of the total mortality to the dosage is shown in Fig. 12.*

If the standard deviation of the log-tolerance distributions for the two poisons is altered, without change of the mean probit

* Points on this curve are most easily calculated by first tabulating Y_1 and Y_2 for a series of values of x; for example, when $x = 0\cdot20$ for the mixture, the log concentration of each constituent present in the mixture is $x = -0\cdot10$, so that $Y_1 = 5\cdot8$, $Y_2 = 3\cdot8$. Values of Q_1, Q_2,

difference, the only change in Fig. 12 is a change in the scale of x.
For example, for two poisons giving the separate lines

$$Y_1 = 6 + 10x, \quad Y_2 = 4 + 10x,$$

a 1 : 1 mixture shows the same response curve as Fig. 12, except
that one unit of x in that figure must now be read as only $\frac{1}{5}$ unit.

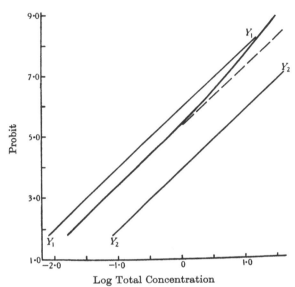

FIG. 12. Independent action in a 1 : 1 mixture of poisons with probit regression
lines $Y_1 = 6 + 2x$, $Y_2 = 4 + 2x$. Curve shows dosage-response relationship for
mixture (Ex. 24). Broken line shows dosage-response relationship for correlated independent action with $r = 1$ (Ex. 26).

If the standard deviation remains unaltered, but the mean
log tolerances are changed, the curve is again unaltered as long
as the relative potency is the same. For example, if

$$Y_1 = 3 + 2x, \quad Y_2 = 1 + 2x,$$

the proportions of test subjects separately surviving these dosages,
are then obtained from Table 1; here $Q_1 = 0.2119$ and $Q_2 = 0.8849$. Now
equation (8·18) can alternatively be written

$$Q = Q_1 Q_2,$$

so that tabulation of products gives the proportion surviving the mixed
dose, or in this instance $Q = 0.1875$; the value of $Y = 5.89$ is then read
directly from Table 1.

the two lines and the curve are shifted horizontally by 1·5 units; the change may be accomplished by writing $(x+1\cdot5)$ for x in Fig. 12.

If the relative potency of the two poisons is changed, the curve for the mixture is changed though still very similar in general form. For the lines

$$Y_1 = 5\cdot1 + 2x, \quad Y_2 = 4\cdot9 + 2x,$$

a 1 : 1 mixture gives the curve in Fig. 13.

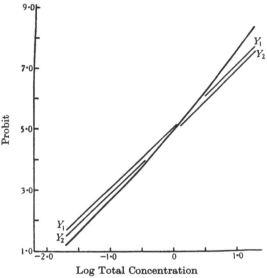

FIG. 13. Independent action in a 1 : 1 mixture of poisons with probit regression lines $Y_1 = 5\cdot1 + 2x$, $Y_2 = 4\cdot9 + 2x$. Curve shows dosage-response relationship for mixture (Ex. 24).

The contrast between similar action and independent action is emphasized by consideration of the interesting limiting case of a mixture of two poisons whose separate probit lines are identical. Under similar action, any such mixture, irrespective of the proportions, would give the same line. Under independent action a curve is obtained, of a form very like those just discussed, which at low concentrations lies below the lines for the two constituents, but at high concentrations lies above. If

$$Y_1 = Y_2 = 5 + 2x,$$

the curve for a 1:1 mixture is obtained by calculating, for various values of x,

$$Q = Q_1^2;$$

this curve is shown in Fig. 14, together with further curves for mixtures of 4, 16 and 1000 components in equal proportions, all having the same individual probit lines.

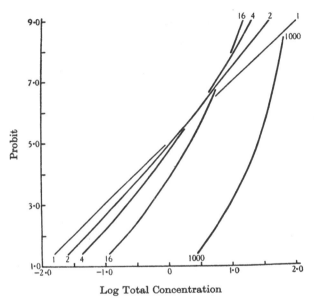

FIG. 14. Independent action in mixtures of several poisons each with probit regression line $Y = 5 + 2x$. Curves are drawn for 2, 4, 16 and 1000 components (Ex. 24).

From consideration of Figs. 12–14 the general nature of the curve for a mixture in equal proportions, at least when the two constituents give parallel probit lines, may be inferred. At low concentrations the curve is indistinguishable from the line representing the effect of the mixture's content of the more potent constituent applied alone; if the relative potency of the more potent constituent is less than two (Fig. 13) the kill at these concentrations is even less than for the same total concentration of the less potent constituent alone. At higher concentrations, the kill is augmented by an appreciable effect of the weaker constituent, so that the curve turns upward and is no longer parallel to the two original lines. If the two constituents do not differ greatly in potency (Figs. 13, 14), as the concentration is increased the mixture soon becomes more effective than the same total concentration of the stronger constituent, but if the difference is large (Fig. 12) this does not occur until very high concentrations and kills are reached.

The discussion just given may be extended to cover mixtures in proportions other than 1 : 1 by a simple transformation. A 2 : 3 mixture, for example, may be considered as a mixture in equal proportion of the first constituent with a new second constituent of 3/2 the potency of the original. Hence on the probit diagram the line for the second constituent may be shifted a distance $b \log (3/2)$ to the left and the behaviour of the mixture then deduced as that appropriate to a 1 : 1 mixture. For numerical calculation nothing is gained by this process, as the ordinates of the curve for the mixture may just as easily be calculated directly.

Bliss (1939 a) implied that mixtures of two poisons showing similar action always had greater potency than mixtures of two poisons with potencies equal to those of the first pair but showing independent action. This is true at moderate dose rates if the relative potency of the constituents is large, but Figs. 13 and 14 indicate that if the relative potency approaches unity independence gives a lower potency at low doses and a higher potency at high doses than does similarity.

Ex. 25. *Independent action between constituents with intersecting probit lines*. The curve relating mortality probit to dosage for a 1 : 1 mixture of independently acting poisons whose separate regression equations are

$$Y_1 = 5 + 2x, \quad Y_2 = 5 + 4x,$$

is shown in Fig. 15; in general characteristics it is typical of the curve resulting from mixing poisons whose regression lines intersect. At concentrations sufficiently low for the constituent with the greater log-tolerance variance (i.e. lesser value of b) to be the more potent, a 1 : 1 mixture produces almost the same effect as would its content of this constituent applied alone. At higher concentrations, the curve for the mixture eventually rises rather more steeply than the line for the constituent with the greater value of b; as the curve increases in steepness with increasing concentration it may be presumed eventually to intersect the line for this constituent alone, but unless the original lines are nearly parallel this intersection seems not to occur until very high kills have been reached.

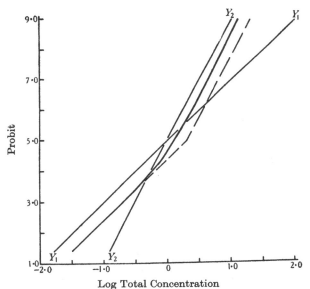

FIG. 15. Independent action in a 1 : 1 mixture of poisons with probit regression lines $Y_1 = 5 + 2x$, $Y_2 = 5 + 4x$. Curve shows dosage-response relationship for mixture (Ex. 25). Broken line shows dosage-response relationship for correlated independent action with $r = 1$ (Ex. 26).

Curves calculated for 1 : 1 mixtures whose constituents have the regression lines

$$Y_1 = 4 + 2x, \quad Y_2 = 4 + 4x,$$

or
$$Y_1 = 2 + 2x, \quad Y_2 = 2 + 4x,$$

have been found to be very like those of Fig. 15, except for the displacement caused by the changed point of intersection of the lines, but they are not quite identical in shape. Two poisons whose probit lines are

$$Y_1 = 5 + 2x, \quad Y_2 = 5 + 10x,$$

differ more markedly in tolerance variance than those just considered, but the curve for a 1:1 mixture has the same general appearance (Fig. 16). Mixtures in proportions other than 1:1 may be discussed as 1:1 mixtures by means of the transformation suggested at the end of Ex. 24.

Ex. 26.   *Correlated independent action.* The curve representing the toxic effect of a mixture of two poisons showing correlated independent action, as defined by equation (8·20), is always

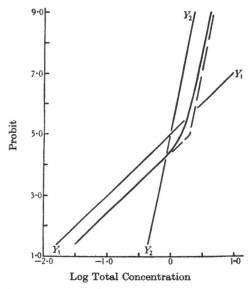

FIG. 16. Independent action in a 1 : 1 mixture of poisons with probit regression lines $Y_1 = 5 + 2x$, $Y_2 = 5 + 10x$. Curve shows dosage-response relationship for mixture (Ex. 25). Broken line shows dosage-response relationship for correlated independent action with $r = 1$ (Ex. 26).

intermediate in position between that for completely independent action and the line or sections of lines for the more potent constituent alone. Thus for complete correlation ($r = 1$) in Ex. 24, since one constituent is always more potent than the other at all dosages, the mixture would follow the line for the amount of this constituent present in the mixture; this line is shown in Fig. 12 and is a continuation of the initial rectilinear portion of the curve for $r = 0$. For less complete correlation, represented by $0 < r < 1$, an intermediate curve would be obtained. In Ex. 25 the more potent constituent is not the same at all dosages. Sections of lines corresponding to $r = 1$ are also shown in Figs. 15 and 16; again lesser values of r would give curves in intermediate positions. The two extremes ($r = 0$ and $r = 1$) are often not very different, and only very extensive and precise data could permit satisfactory discrimination between different values of r. With experimental results for poisons whose effects are like those shown in Fig. 16, for example, it would be very difficult to assess r with any precision. Indeed, in practice, data which are appreciably less satisfactorily fitted by $r = 0$ than by some other value of r will rarely be encountered.

As stated above, no exact statistical treatment of data relating to independent action has yet been developed. If probit lines for the constituents are estimated in the same series of tests as those on the mixture, these lines may be used to predict the form of the curve for the mixture, and the observed mortalities may then be compared with the expected, by a χ^2 test in the manner of that used in Ex. 1. Such a test ignores the errors of estimations of the lines for the constituents, and is therefore liable to exaggerate the significance of discrepancies, but no alternative can at present be suggested. No examples from experimental data will be given here, as none suitable have been found in the published literature.

Mortality probits corresponding to low concentrations of a poison have often been found to be higher than predicted by the line fitted to the whole data. Bliss (1939a) has suggested that this phenomenon of a 'break' in the line may indicate that the poison is in fact a mixture of two or more toxic components; the possibility merits further consideration, but detailed experimentation at many concentrations is needed if it is to be examined adequately. Such breaks are often drawn as sudden changes in slope, a situation which can only occur with independent action if $r = 1$. The data are seldom sufficiently precise, however, for any certainty that the change in slope is not more gradual and of the type found for lesser values of r, perhaps even for $r = 0$. Frequently only the upper portion of the curve is of interest, since this usually extends over the important range of kills, and difficulties of analysis may be avoided by ignoring the results for the lower dosages, a policy which is theoretically objectionable but practically justifiable.

Murray (1938) published the results of tests on the toxicity of a pyrethrin spray to the house-fly which give some evidence of independent action of pyrethrins I and II; a distinct 'break' occurs for the female flies, but the more susceptible males show only one phase of action over the whole range of concentrations tested. The proportions of the two pyrethrins present in the spray were not stated, nor were tests made on either component alone, so that no critical judgement can be formed. The data have been discussed by Bliss.

39. SYNERGISTIC ACTION

Many writers have used the terms *synergism* and *antagonism* to describe the joint action of certain mixtures of poisons without giving any unambiguous definition of their meaning. The first attempt to formulate algebraic relationships which would represent this type of action was due to Bliss (1939*a*); he proposed two alternative equations for the simple case of *similar synergistic action*, in which the probit-dosage regression lines for mixtures and for the constituent poisons are all parallel.

Bliss endeavoured to find the relationship between the quantities of two poisons present in equipotent doses of mixtures behaving synergistically. If a total dose λ contains quantities λ_1 and λ_2 of two constituents, the first being the 'more active ingredient', an equation will connect the values of λ_1 and λ_2 for which the total kill is constant.* The first equation suggested by Bliss may be written in the form

$$\lambda_1^\epsilon(\lambda_1 + \lambda_2) = k, \qquad (8\cdot21)$$

where ϵ depends only upon the two poisons and k is a function only of the level of kill selected. Bliss does not state by what criterion it is to be decided which ingredient is the more active, though some standard independent of dose seems to be implied.

* For similar action, the equation is $\lambda_1 + \rho\lambda_2 = $ constant.

The equation cannot hold for very small doses of the first poison, since it implies that the addition of a trace of this to a moderate dose of the second would reduce rather than increase the potency.

The second equation, which Bliss considered more satisfactory, is

$$\lambda_2^\epsilon(1 + h\lambda_1) = k, \qquad (8\cdot22)$$

where ϵ again depends only on the two poisons and h, k are functions of the level of kill selected for comparing the data. On the analogy of Clark's findings for drug antagonism (1937), Bliss suggests that ϵ is frequently very close to unity; he also states that in one example the product hk appeared to be almost independent of the kill and proposes that its value might therefore be used as a measure of the intensity of synergism. Equation (8·22) must break down at very small doses of the second poison since it implies that a preparation containing only a trace of the second mixed with the first would be less toxic than the same amount of the first applied alone.

Both these equations must be judged unsatisfactory representations of similar synergistic action. Apart from the discontinuity for small doses and the difficulty of defining the more active ingredient, there is the disadvantage that neither includes similarity or independence, the two simpler forms of joint action already discussed, as cases of zero synergism. Though the final appeal must be to experiment rather than to abstract argument, equations (8·21) and (8·22) appear unlikely to be very helpful in describing synergistic action, particularly as each requires at least two parameters, one of which depends upon the level of mortality.

An alternative equation, which may satisfactorily fit some data whose probit regression equations, both for mixtures and for their constituents, are rectilinear and parallel, has been suggested by Finney (1942a). It may be written, using the same notation as in § 36,

$$Y = a + b \log (\pi_1 + \rho\pi_2 + \kappa \sqrt{[\rho\pi_1\pi_2]}) + bx, \qquad (8\cdot23)$$

and is an extension of equation (8·9). A dose λ of the mixture produces the same effect as a dose $(\pi_1 + \rho\pi_2 + \kappa \sqrt{[\rho\pi_1\pi_2]})\lambda$ of the first constituent alone. If $\kappa = 0$, equation (8·23) represents similar action. If κ is positive, the potency is greater than that predicted by similar action, and the poisons act synergistically; if κ is negative, they act antagonistically. Equation (8·23) may prove at least a satisfactory empirical formula for mixtures whose probit lines are parallel to those for their constituents; the constant κ will be called the *coefficient of synergism*. The equation will easily generalize so as to include mixtures of several toxic constituents.

In order that the agreement between experiment and a proposed law of synergistic action may be examined, toxicity tests must be carried out with a range of doses of mixtures in at least two different proportions as well as on the constituents separately. From data collected in such an experiment the coefficient of synergism, as defined above, can be estimated by a generalization of the method described in § 37, using now a regression equation in the form

$$\frac{1}{\lambda} = \left(\frac{1}{\lambda_1}\right)\pi_1 + \left(\frac{\rho}{\lambda_1}\right)\pi_2 + \left(\frac{\kappa\rho^{\frac{1}{2}}}{\lambda_1}\right)(\pi_1\pi_2)^{\frac{1}{2}}. \qquad (8\cdot24)$$

The only data found in published papers which give definite evidence of synergism and which are shown in a form suitable for testing the adequacy of equation (8·23) are discussed in Ex. 27. Before the many problems of synergism can be elucidated, a great amount of further experimentation, testing mixtures in several different proportions, must be carefully planned and executed. Without much more experimental evidence than is at present available any complete discussion even of similar synergistic action is impossible; no attempt will be made here to unravel the complexities that may arise when the separate regression lines are not parallel.

Ex. 27.   *The toxicity of rotenone-pyrethrins mixtures to the house-fly.* Le Pelley and Sullivan (1936) have reported the results of two series of trials in which adult house-flies were sprayed with alcoholic solutions of rotenone, pyrethrins, and a mixture of the two; in the first series the mixture contained rotenone and pyrethrins in the proportion of 1:5 (by weight), and in the second series the proportion was 1:15. About 1000 flies were tested at five levels of each toxic preparation, and the kills obtained were as shown in Table 25, these figures having been abstracted from the diagrams in the original paper. The authors interpreted the results as indicating no striking antagonistic or synergistic effect in the mixtures. A footnote by H. H. Richardson asserted that in the first series there was pronounced synergism and in the second an effect in the same direction but of less

TABLE 25. Toxicity of Rotenone and Pyrethrins to House-flies

First series			Second series		
Concentration (mg./c.c.)	No. of flies	% kill	Concentration (mg./c.c.)	No. of flies	% kill
Rotenone			Rotenone		
0·10	1000	24	0·10	900	28
0·15	1000	44	0·15	900	51
0·20	1000	63	0·20	900	72
0·25	1000	81	0·25	900	82
0·35	1000	90	0·35	900	89
Pyrethrins			Pyrethrins		
0·50	1000	20	0·50	900	23
0·75	1000	35	0·75	900	44
1·00	1000	53	1·00	900	55
1·50	1000	80	1·50	900	72
2·00	1000	88	2·00	900	90
Mixture (1 : 5)			Mixture (1 : 15)		
0·30	1000	27	0·40	900	23
0·45	1000	53	0·60	900	48
0·60	1000	64	0·80	900	61
0·875	1000	82	1·20	900	76
1·175	1000	93	1·60	900	93

apparent significance; Richardson here used a prediction for the mixture equivalent to the similar action law. Bliss (1939a) confirmed Richardson's conclusion, and Finney (1942b), after a new analysis of the data, also agreed that there was evidence of synergism.

In this last analysis, for each series, parallel probit lines were fitted to the data for the two poisons and their mixture. For the first series, the LD 50's were 0·156, 0·918 and 0·455 mg./c.c. for rotenone, pyrethrins and the 1 : 5 mixture respectively. The pyrethrins were slightly more than one-sixth as toxic as the rotenone (more precisely, $\rho = 0·170$), whence the similar action law predicts a relative potency of 0·308 for the mixture, or an LD50 of 0·506 mg./c.c. The mixture was thus 11 % more potent than would be expected if its constituents acted similarly; a test of the significance of this synergistic effect may be obtained from

$$\Delta_s = 0\cdot168 \pm 0\cdot050,$$

leaving little doubt that the enhanced toxicity is greater than can be attributed to random sampling variation. Analysis of the second series gave 0·142, 0·889 and 0·651 mg./c.c. as the LD50's for rotenone, pyrethrins and the 1:15 mixture. The values for the two constituents of the mixture were thus very close to those obtained in the first series, and gave $\rho = 0\cdot160$. Similar action then predicts a relative potency of 0·212 for the mixture, and therefore an LD50 of 0·670 mg./c.c. This mixture was 3 % more potent than predicted, but the difference here is not significant since

$$\Delta_s = 0\cdot039 \pm 0\cdot067.$$

Both series of tests give some indication of synergism, though in only one is the departure from similarity significant. It is therefore of interest to inquire whether the equation for similar synergistic action (equation (8·23)) will fit the results. The data are insufficient for any precise estimation of the coefficient of synergism, but a value of about 0·15 may easily be seen to be satisfactory. Using $\kappa = 0\cdot15$, the expression $(\pi_1 + \rho\pi_2 + \kappa\sqrt{[\rho\pi_1\pi_2]})$ gives the expected potency of a mixture relative to that of rotenone; for the two mixtures under test the values are 0·331 and 0·227 respectively. The corresponding LD50's, 0·471 and 0·626 mg./c.c., agree well with the experimental determinations of 0·455 and 0·651 mg./c.c.; even without exact statistical tests the discrepancies are seen to be not significant, so that at least the data do not contradict the hypothesis expressed by equation (8·23).

40. PLANNED TESTS OF SYNERGISM AND SIMILARITY

When experiments on the joint action of two poisons have to be planned in the absence of any information about the existence of synergism or antagonism between them, a working hypothesis that the action will be similar may reasonably be adopted. Unless there are *a priori* considerations governing the proportionate constitutions of mixtures to be tested, it is preferable that the probit lines for mixtures should be fairly evenly spaced between those for the constituents used separately. If the second constituent is ρ times as potent as the first, its probit line will be a distance $\log \rho$ to the left of the first. On the hypothesis of similar action, if a mixture in the proportion $\pi : (1 - \pi)$ yields a probit regression line at a distance $\theta \log \rho$ to the left of that for the first poison, where θ is some fraction between 0 and 1, then

$$\log \{\pi + \rho(1 - \pi)\} = \theta \log \rho,$$

whence
$$\pi = \frac{\rho - \rho^{\theta}}{\rho - 1}. \qquad (8 \cdot 25)$$

In Table 26 π is tabulated, as a percentage, for a series of values of ρ and θ; providing that an approximation to ρ is available from earlier experiments, the table may conveniently be used in planning toxicity tests intended for the investigation of similar and similar synergistic action.

Ex. 28.   *The use of Table 26 in planning toxicity tests.* Suppose that toxicity tests are to be planned for two poisons and a mixture whose constitution shall be chosen so that, if similar action is operating, the probit regression line for the mixture will be midway between the lines for the constituents, all dosages being measured as the logarithms of total poison content. Suppose further, that, from previous experience, the second poison is believed to be about four times as toxic as the first. Entering Table 26 with $\rho = 4$ in the column for $\theta = 0 \cdot 5$, the required proportion of the first poison is found to be 67 %, and therefore a 2 : 1 mixture of the two poisons should be used.

Again, suppose that three different mixtures of two similarly acting poisons are required such that the mixtures give probit lines equally spaced between those for the separate poisons, and assume that the second constituent is known to be about twelve times as potent as the first. Interpolation in Table 26 for $\rho = 12$ and $\theta = 0 \cdot 25$, $0 \cdot 5$ and $0 \cdot 75$ gives figures of about 91, 77 and 50 %

for the amounts of the first poison in the mixtures. Suitable mixtures would therefore be made in the proportions of $10:1$, $7:2$ and $1:1$.

In order to obtain reliable evidence of the nature of the joint action of the two constituents of a mixture, a minimum of four concentrations of each toxic preparation should be tested. If

TABLE 26. The Function $\pi = 100(\rho - \rho^\theta)/(\rho - 1)$ used in Planning Tests of Mixtures of Two Poisons

ρ \ θ	0·1	0·2	0·3	0·4	0·5	0·6	0·7	0·8	0·9
1·1	90·4	81	71	61	51	41	31	21	10
1·5	91·7	83	74	65	55	45	34	23	12
2	92·8	85	77	68	59	48	38	26	13
3	94·2	88	80	72	63	53	42	30	16
4	95·0	89	83	75	67	57	45	32	17
5	95·6	90·5	84	77	69	59	48	34	19
6	96·1	91·4	86	79	71	61	50	36	20
7	96·4	92·1	87	80	73	63	52	38	21
8	96·7	92·6	88	81	74	65	53	39	21
9	96·9	93·1	88	82	75	66	54	40	22
10	97·1	93·5	89	83	76	67	55	41	23
15	97·8	94·9	91·0	86	79	71	60	45	25
20	98·2	95·7	92·3	88	82	74	62	47	27
25	98·4	96·2	93·2	89	83	75	65	49	29
30	98·6	96·6	93·9	90	85	77	66	51	30
40	98·9	97·2	94·8	91·3	86	79	69	54	32
50	99·0	97·6	95·4	92·3	88	81	70	55	33
60	99·1	97·9	95·9	93·0	89	82	72	57	34
80	99·3	98·2	96·6	94·0	90	84	74	59	36
100	99·4	98·5	97·0	94·6	90·9	85	76	61	37

there is no prior knowledge of synergism, the test doses of each should roughly be inversely proportional to their relative potencies. For example, consider the trials on two poisons of relative potency twelve and their three mixtures in the proportions determined in the last paragraph. The relative potencies of the five preparations are, by equation (8·5), approximately $2:4:7:13:24$, and doses inversely proportional to these numbers should be used. Hence if the LD 50 for the stronger poison were thought to be about 0·2 unit and experience had shown the suitability of a two-fold increase in dose between successive levels, the following sets of five doses of each preparation might be chosen:

Mixture in proportions of

	1:0	10:1	7:2	1:1	0:1
	0·004	0·008	0·015	0·025	0·05
	0·008	0·016	0·03	0·05	0·1
Doses	0·016	0·032	0·06	0·1	0·2
	0·032	0·064	0·12	0·2	0·4
	0·064	0·13	0·24	0·4	0·8

41. COMPOUND RESPONSE CURVES

An unusual dosage-response curve has been reported (Dimond *et al.* 1941) for tests of the toxic effect of tetramethylthiuram disulphide to spores of *Macrosporium sarcinaeforme*. Dr Dimond has kindly made available more information about these tests than was originally published. Five concentrations, ranging from 0·2 % to 0·00002 %, were sprayed on to glass slides, and several different spray times between 50 and 5 sec. were tested for each concentration. In this way, deposits ranging from 35 to 0·00105 μg./sq.cm. were obtained, some of these being duplicated by being made up from two different combinations of concentration and time. Spore suspensions were then added to the dried residues of the spray so that the concentration of the toxicant in the drop of suspension was proportional to the density of the dried deposit on the slide. The percentage inhibition of spore germination was measured; Fig. 17, which has been copied from that published by Dimond *et al.*, shows the relationship between the inhibition probits and log deposits. The very close agreement between percentage inhibitions from pairs of results with equal deposits, but different combinations of duration of spraying and concentration, suggests that the total deposit was the chief factor determining the response and that the method of obtaining this deposit was comparatively unimportant. The curve indicates that tetramethylthiuram disulphide had a maximum potency at about 0·06 μg./sq.cm., above which level any increase in dose decreased the inhibition of germination until a minimum was

reached at about 0·3 µg./sq.cm. At still higher doses, the potency
again increased and rose beyond the previous maximum; inhibition was complete for the six highest levels tested.

The authors make the following comments:

'A possible explanation of this type of behaviour is that dissociation or association of the toxicant occurs. It seems likely
that weakly-dissociating materials may dissociate (or associate),

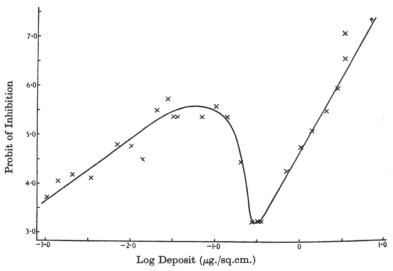

FIG. 17. Toxicity of tetramethylthiuram disulphide to spores of *M. sarcinaeforme*. Each point is based on a count of 100 spores. Complete inhibition was
recorded for log deposits of 0·85, 1·02, 1·15, 1·32, 1·45 and 1·54.

forming a complex which has toxicity markedly different from the
original molecule. In the case of tetramethylthiuram disulfide,
the toxicity of the dissociation complex would be greater than
that of the undissociated molecule.

'In the first phase of toxic action,* where decrease in toxicity
is proportional to dilution on the logarithmic-probability scale,
the proportion of dissociated molecules as compared with the
dissolved, undissociated molecules of toxicant might be very

* The authors explain that the argument in this paragraph has been
put in terms of decreasing concentration instead of increasing, in order
to simplify the discussion of the effect of molecular dissociation.

small, and toxic action would be determined largely by the undissociated molecules. As dilution continues, however, the proportion of dissociated to undissociated molecules increases, and, if the dissociated molecule is very much more toxic than the undissociated molecule, the inhibition of spores will rise with further dilution. Finally, when the original toxicant is completely dissolved and dilution has progressed to the stage at which there are no undissociated molecules left, the second peak of toxicity has been reached. Dilution beyond this point can only cause a decrease in concentration of dissociated molecules and in toxicity.'

Thus they suggest that tetramethylthiuram disulphide has two separate types of toxic action, that of the undissociated molecule occurring when high concentrations are used, and that of the dissociated molecule occurring when low concentrations are used. At intermediate concentrations presumably the toxicant must behave as a mixture of the two toxic materials; the proportionate constitution of the mixture will depend upon the total concentration of the poison, and as the concentration decreases the proportion of the dissociated molecule will change smoothly from 0 to 1. Montgomery and Shaw (1943) have confirmed the occurrence of this form of response curves for several thiuram sulphides with spores of *Venturia inaequalis*.

Any study of the form of dosage-response curve that may be expected in these circumstances requires knowledge not only of the potency of a mixture of the two toxic constituents in any given proportions but also of the law determining the proportions that occur at any concentration. The two extreme linear sections of the curve in Fig. 17 are clearly not parallel, and therefore similar action cannot be operative. But, even with parallel lines to represent the extreme phases of toxic action, an increase in toxicity with decrease in concentration at intermediate levels could not occur as a result of similar action unless the total concentration of dissociated molecules, not merely the proportion of dissociated to undissociated, were then increasing; the dissociation would therefore have to be of a different type from that normally encountered, since this could not happen according to accepted dissociation laws. On the other hand, if there were a pronounced antagonism between the toxic action of the dissociated and undissociated molecules, a curve similar to Fig. 17 might arise. Thus a possible explanation of Dimond's results is that the dissociated and undissociated molecules differ widely in their potency and in the tolerance variance shown by the spores (as evidenced by the difference in slope of the two linear sections), and that some type of antagonism is displayed when the two are in mixture. Further understanding of this complex problem must await the accumulation of a greater amount of experimental evidence.

As an attempt to set up a mathematical model of what might occur when two toxic materials which show similar action in mixture are mixed, under conditions that make the proportionate constitution of the mixture dependent upon the total concentration, a study has been made of the behaviour of a mixture in proportions

$$\pi_1 = 1 - \pi_2 = 10^{-k\lambda}.$$

Here λ is the total concentration of poison, k a positive constant; the law is not intended as even an approximation to any physico-chemical relationship, but has simply been chosen so that the proportion of the first constituent, π_1, is zero at high concentrations and changes continuously to 1 at low concentrations. If

$$Y = a + b \log \lambda$$

gives the probit of the kill for this constituent, and the relative potency of the second is ρ, equation (8·4) gives the probit of the kill for any concentration as

$$Y = a + b \log \{\rho + (1 - \rho) \, 10^{-k\lambda}\} + b \log \lambda. \qquad (8·26)$$

If ρ is less than $1/(e^2 + 1)$, or about 0·119, equation (8·26) has both a maximum and a minimum, and represents a curve of the same general form as that in Fig. 17, except that the two extreme sections are now parallel.

Ex. 29. *Response curves given by equation* (8·26). Fig. 18 shows the curve derived from equation (8·26) when the probit line for the first constituent is

$$Y = 5 + 2 \log \lambda,$$

the relative potency of the second constituent is $\frac{1}{100}$, and $k = \frac{1}{20}$. These values have been chosen as roughly corresponding to the curve obtained by Dimond *et al.*; the similarity with Fig. 17 is noticeable, especially in respect of the wide peak and narrow trough, though the trough in Fig. 17 is deeper. A modification of equation (8·26) so as to base it on similar synergistic action would alter the proportions of the curve in accordance with the

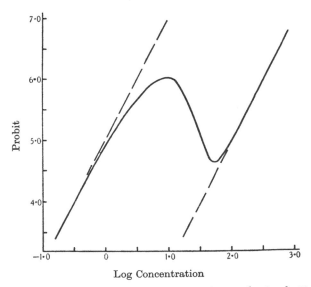

FIG. 18. Response curve from equation (8·26) for $\rho = \frac{1}{100}$, $k = \frac{1}{20}$ (Ex. 29).

amount of synergism or antagonism, but unless the departure from similarity were very great the main features would remain.

When ρ is greater than 0·119 but less than 1, the maximum and minimum disappear, and increasing concentration always increases the kill. Fig. 19 has been drawn for $\rho = 10^{-\frac{1}{4}}$, $k = \frac{1}{20}$, the probit line for the first constituent being the same as before. Typical of the situation in which the second constituent is the more toxic is Fig. 20; here $\rho = 100$, $k = \frac{1}{20}$, and the equation for the first constituent is again the same. This curve has been drawn for probits far beyond the range of practical importance, in order to show its eventual agreement with the upper line.

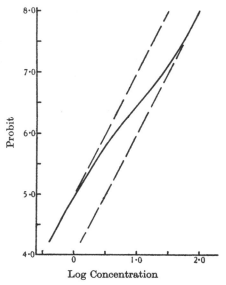

FIG. 19. Response curve from equation (8·26) for $\rho = 10^{-\frac{1}{2}}$, $k = \frac{1}{2 \cdot 0}$ (Ex. 29).

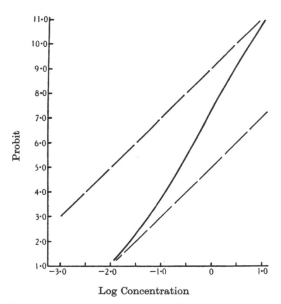

FIG. 20. Response curve from equation (8·26) for $\rho = 100$, $k = \frac{1}{2 \cdot 0}$ (Ex. 29).

Further discussion of the possible effects on the dose-response curve of molecular dissociation of poisons or of other types of compound response curves would be unprofitable at this stage. The only conclusion to be drawn at present is that the data of Dimond *et al.* are in general agreement with the behaviour that might occur if undissociated and dissociated molecules have very different toxicities and behave antagonistically. The account that has been given in this section is intended only as a suggestion of a mode of action that may sometimes be found, and is in no way an exact treatment of this complex problem; more detailed consideration is impossible without a much greater amount of experimental evidence as well as a fuller understanding of the underlying theory.

Chapter 9

MISCELLANEOUS PROBLEMS

42. VARIATION BETWEEN BATCHES

IN the analyses discussed in previous chapters it has been tacitly assumed that all the subjects at one dose were members of a single batch and were tested at the same time. Many assays of insecticides and fungicides are in fact conducted by exposing several distinct batches to each dose of the poison. For example, in some forms of apparatus for the testing of insecticidal sprays not more than about twenty insects can be used at one time, and therefore, if conclusions based on approximately sixty insects at each dose are required, three batches must be tested. Examination of the variability between mortalities in batches given the same dose then provides a measure of the heterogeneity of the behaviour of the batches, including both biological differences between batches and variation from batch to batch in experimental technique. Comparison of this variability with the residual variation between doses (after the removal of the probit regression component) enables a test to be made of whether the latter can be explained as due only to the natural batch variation, and thus of whether the regression line is an adequate description of the relationship between dosage and response.

The usual procedure when several batches are used for each dose is merely to add the values of n (number of test subjects) and r (number killed) and to treat the results as if they referred to a single batch. Very often this course is satisfactory. If all batches contain the same number of subjects, the maximum likelihood estimates of the parameters are the same whether all subjects for one dose are treated as a single batch or the batches are kept distinct throughout, and, providing that the same provisional line is used, the estimates are the same at the end of each cycle of computations. This remains true for batches of unequal size as long as the total variance of the percentage mortality at a given dose is inversely proportional to n. Even when there is a component of batch variance independent of n, the batches will often be of nearly the same size, and the less onerous computations needed when batches are combined make that the generally preferred method.

When the residual variation between doses after the fitting of the linear probit regression equation gives a significant χ^2, this may indicate either a real departure from linearity or a non-independence of the responses of individuals of the same batch. ·If the latter explanation is adopted, all variances have to be multiplied by a heterogeneity factor, and, since the degrees of freedom available for the estimation of this factor are often very few, its estimate may be of low precision; fiducial limits will then vary irregularly in repeated determinations and will tend to be widely spaced on account of the increased value of t for few degrees of freedom. By using the data for each batch separately, not only may a test of linearity be derived but, supposing that there is heterogeneity between batches though no significant departure from linearity, a heterogeneity factor based on a greater number of degrees of freedom may be estimated.

At this point a brief digression on the order in which a series of tests should be carried out seems appropriate. Strictly speaking, the different batches of test subjects should be assigned to the doses entirely at random and the doses should be tested in random order, or in some restricted randomization in accordance with modern principles of experimental design (Fisher, 1942; Yates, 1937a). In practice, order of testing is usually held to be unimportant in a well-controlled experiment, provided that the whole is completed within a reasonably short time (with insecticides, preferably within one day); the theoretical requirements of randomization are therefore frequently sacrificed to the practical convenience of testing all batches at one dose consecutively and making all tests on one poison before those on a second are begun. The possibility of bias from this source must not be neglected, and every precaution to eliminate it must be taken lest any difference in susceptibility of insects tested at different times in the day be interpreted as a difference in potency of two poisons (McLeod, 1944). If the doses of a single poison are tested in ascending or in descending order, any steady trend in susceptibility of the subjects or any increase in the dose actually received (through incomplete cleaning of the apparatus after the preceding dose) may manifest itself by an increase or decrease in the estimated slope of the regression line or even by a departure from linearity; if the doses are given in random order, though these influences may increase the heterogeneity between batches they will not bias any estimates of potency. In insecticidal work the random selection of insects for the batches may be important; any method of selection which allows the more active insects to be used first, or which takes insects from a culture as they reach a certain stage of development, may have undesirable consequences through the batches being heterogeneous in respect of sex ratio or some other important factor correlated with susceptibility to the poisons under test (Murray, 1937; Bliss, 1939b). In other branches of biological assay analogous considerations arise, and, unless a strict randomization has been employed throughout, the experimenter should always be on his guard against any bias in his data resulting from non-random selection and order of testing.

When there is no evidence of heterogeneity of any type—and, with experience, this may often be judged from inspection of the data before any statistical analysis is made—there is nothing to be gained by maintaining the identities of different batches tested at each dose, and the less laborious analysis of combined batches should be adopted. When there is heterogeneity, such as is indicated, in the usual manner, by a significant residual χ^2, a complete analysis may sometimes be made with advantage, in order to gain degrees of freedom for the heterogeneity factor, even though the estimates of potency will not be affected. If more than one cycle of computations is likely to be needed the detailed work should only be done in the last cycle, since the only use of the earlier ones is to improve the provisional line. The tests to be made can best be described by means of a numerical example.

Ex. 30. *The toxicity of ammonia to* Tribolium confusum. Strand (1930, Table I) has given the results of tests of ammonia as a fumigant for *T. confusum*. Two batches of insects were tested at each of eight concentrations of the fumigant; from the form in which Strand states the results it seems likely that the tests were carried out as two distinct experiments, one batch in each, but for present purposes this point will be ignored and the pairs of batches will be assumed to be replicates of the type just discussed.

The computations necessary for fitting a probit regression line, when the batches are all kept separate, are shown in Table 27.

TABLE 27. Computations for Ammonia-*Tribolium confusum* Tests

x	n	r	p	Empirical probit	Y	nw	y	nwx	nwy	
0·72	29	2	7	3·52	3·3	6·0	3·57	4·320	21·420	40·260
	29	1	3	3·12		6·0	3·14	4·320	18·840	
0·80	30	7	23	4·26	4·1	14·1	4·27	11·280	60·207	131·309
	31	12	39	4·72		14·6	4·87	11·680	71·102	
0·87	31	12	39	4·72	4·8	19·5	4·72	16·965	92·040	173·043
	32	4	12	3·82		20·1	4·03	17·487	81·003	
0·93	28	19	68	5·47	5·5	16·3	5·47	15·159	89·161	182·401
	31	18	58	5·20		18·0	5·18	16·740	93·240	
0·98	26	24	92	6·41	6·0	11·4	6·32	11·172	72·048	151·880
	31	25	81	5·88		13·6	5·87	13·328	79·832	
1·02	27	27	100	∞	6·4	8·2	6·94	8·364	56·908	113·603
	28	27	96	6·75		8·5	6·67	8·670	56·695	
1·07	26	26	100	∞	6·9	4·0	7·34	4·280	29·360	60·176
	31	29	94	6·55		4·8	6·42	5·136	30·816	
1·10	30	30	100	∞	7·2	2·8	7·59	3·080	21·252	40·152
	31	30	97	6·88		2·8	6·75	3·080	18·900	
						170·7		155·061	892·824	

$1/Snw = 0\cdot00585823, \quad \bar{x} = 0\cdot9084, \quad \bar{y} = 5\cdot2304.$

$Snwx^2$	$Snwxy$	$Snwy^2$
142·46391	827·4877	4869·816
140·85480	811·0263	4669·799
1·60911	16·4614	200·017
		168·402
		31·615

They follow the usual plan (cf. Table 6), but, for use at a later stage, subtotals for each dosage are shown in the column nwy. In the table x is the logarithm of the ammonia concentration (in mg./l.) and the other symbols have their usual meanings. The residual sum of squares, 31·62, instead of being tested simply as a $\chi^2_{[14]}$ can be subdivided into two portions. The subdivision is accomplished by first calculating a sum of squares for y between dose levels, the expression for which is

$$(40\cdot260)^2/12\cdot0 + (131\cdot309)^2/28\cdot7 + \ldots + (40\cdot152)^2/5\cdot6$$
$$- (892\cdot824)^2/170\cdot7$$
$$= 4856\cdot852 - 4669\cdot799$$
$$= 187\cdot053;$$

the numerators of the fractions are the squares of the subtotals in the nwy column and the denominators the subtotals of nw for the doses. The analysis of variance of the mortality probits can now be completed as in Table 28. The first line is the sum of squares removed by the regression, the third line is the total sum of squares between the eight doses, the fifth line is S_{yy}, and the other lines are obtained by subtractions. The sums of squares in the second and fourth lines add to give the previously calculated residual, 31·62. The sum of squares in the fourth line is a $\chi^2_{[8]}$ which gives a test of the homogeneity of the results for different batches; its non-significance indicates that any heterogeneity is not sufficiently great to be disclosed by inter-batch variations. The sum of squares which measures departures from linearity, 18·65, if tested as a $\chi^2_{[6]}$, is judged significant, but inspection of the probit diagram (not shown here) discloses no systematic deviation from linearity. Since the two mean squares are not significantly different (Fisher and Yates, 1943, Table V) but both are greater than unity, their expectation for homogeneous data, the most reasonable conclusion to draw seems to be that heterogeneity between batches has increased both. This heterogeneity may be measured by combining the sums of squares to give a factor of $31\cdot62/14 = 2\cdot26$, with 14 degrees of freedom.

If the analysis had been made only on the totals for the eight doses, without using data from separate batches, only the first three lines of Table 28 would have appeared; in fact, 18·65 would then have arisen in the ordinary way as the residual χ^2, and, since the probit diagram indicated no systematic non-linearity, its significance would have been interpreted as due to heterogeneity between batches. The heterogeneity factor would then have been taken as 3·11 with only 6 degrees of freedom. In either analysis, the regression equation is obtained as

$$Y = -4·063 + 10·230x,$$

and other parameters are estimated as required.

TABLE 28. Analysis of Variance of Mortality Probits for Ammonia-*Tribolium confusum* Tests

	D.F.	Sum of squares	Mean square
Regression	1	168·40	
Deviations from linearity	6	18·65	3·11
Between doses	7	187·05	
Between batches at a dose	8	12·97	1·62
Total	15	200·02	

Even when the number of batches is not the same for every dose the calculations follow exactly the same plan. If three or more batches are tested at each dose the number of degrees of freedom for differences between batches will be substantially greater than for deviations from linearity, thus enabling a more sensitive test of heterogeneity to be made and leading to a more precise estimate of the heterogeneity factor. In addition, a test for departures from linearity independent of that for heterogeneity can always be made, but, unless the probit diagram indicates systematic non-linearity, any significance in the linearity test should be interpreted as additional evidence of batch heterogeneity.

43. INDIVIDUAL MORTALITY RECORDS

In testing the effect of a drug or poison, it is sometimes impossible to feed the test subjects with predetermined doses; though the doses can only be roughly controlled, however, exact measurement of the amount taken may be possible afterwards. For example, some techniques for the testing of stomach insecticides, such as Campbell's poison-sandwich technique (Campbell and Filmer, 1929; Campbell, 1930), involve the feeding of separate amounts of poison to each insect, though the dose can only be measured after ingestion. Consequently the experimental results consist of a series of doses with, for each, a record of whether or not a single insect was killed. Such techniques tend to be more troublesome to carry out than those commonly employed for testing batches of insects at selected doses, and only comparatively short series are usually obtained. Tests on less than fifty insects in all cannot be expected to yield results of a precision comparable with that obtained when batches of this size are tested at each of several dosages. Nevertheless, the median lethal dose and other parameters can still be estimated by the method of probit analysis. Little modification of the instructions given in preceding chapters is required, but, as the analysis presents some unusual features, a brief account of it will be given. Bliss (1938) has given a similar but fuller discussion; his conclusions in respect of precision, however, must be treated with some reserve.

The first novelty lies in obtaining the provisional regression line necessary for the initiation of the computations. The observations show either zero or 100 % kill for each dose, and cannot therefore be plotted directly as probits. Instead, the results for a set of consecutive doses must be grouped so as to give percentage kills based on ten or more individuals, and the probits of these must be plotted against the mean log dose or other measure of dosage. In some experiments the doses themselves may indicate a convenient grouping, the experimenter having deliberately aimed at certain values. In others the doses may be spread fairly uniformly over the whole range so that the grouping has to be entirely arbitrary; non-independent overlapping groups of about ten might then be used in order to obtain more points, taking, say, the first to the tenth dosages as one group, the sixth to the fifteenth as a second, the eleventh to the twentieth as a third, and so on. A diagram showing the empirical mortality probit of each group plotted against the mean dosage allows a provisional line to be drawn in the usual manner; the grouped data tend to underestimate the slope of the line, which should therefore be drawn so as apparently to err slightly on the side of steepness.

From this stage, the improvement of the line may proceed exactly as described in § 17, each experimental dosage being used individually and each giving a maximum or a minimum working probit according as the corresponding insect was killed or survived. As usual, if the calculated line differs markedly from the provisional it should be used as a new 'provisional' for a second cycle of computations. When thirty or more doses have been tested, the process of working with 'batches of one' is tedious and seldom worth all the trouble involved. Grouping into small independent (non-overlapping) groups of four or five consecutive dosages greatly reduces the labour of finding sums of squares and products, and, though the regression coefficient will be slightly underestimated, the estimate of the log LD 50 will not be greatly affected provided that the experimental dosages are well distributed on either side of it. Difficulties of estimating a provisional line satisfactorily from meagre data frequently make it necessary to carry out two or more cycles of the computations, and, even though it may be intended to base the final line on the analysis of individuals, grouping may assist the rapid completion of the first cycles.

Unfortunately the troubles encountered with the routine χ^2 test for homogeneity (§ 18) arise in their most acute form in an analysis of individual mortality records, and are also severe when very small groups are used. A single instance of survival at a high dose or death at a low may inflate the value of χ^2 to an undue extent, as may easily be seen by considering, for example, the contribution to this χ^2 given by an observation of 1 in a class whose expectation is only 0·05. The sampling distribution of such a χ^2 is very different from that tabulated in Table VI, values usually being much lower than for a true χ^2 but there being also an excess of high values. If valid conclusions are to be reached by means of an ordinary χ^2 test, the method suggested in § 18 must be followed; after the calculation of expectations for each dosage these must be further grouped so as to give reasonably large total expectations, both of dead and alive, in every group, and χ^2 must then be recalculated from these groups. If the total number of insects in the experiment is under fifty, the number of groups that can be formed will be small and the resulting test of homogeneity will not be very sensitive. On the other hand, no easily applied test of greater sensitivity is available. Indeed, even if the individual tolerances of these few individuals were measured directly, a significance test on the departure from normality of the distribution of their logarithms could not be very sensitive; when the results are only obtainable as quantal responses a sensitive test is even less to be expected.

Bliss (1938) proposed to modify the usual test of heterogeneity by referring χ^2, calculated according to equation (4·2), to a distribution with less than $(k-2)$ degrees of freedom (where k is the number of groups). His rule involves classing as one group enough of the terminal groups at each end of the range of doses to make the expected number of survivors at the upper end and deaths at the lower end at least 10 % of the number of subjects in the standard group, even when the latter consists of only one: 'Thus if there were three in a standard group, the end groups at the upper end would be combined until 0·3 or more live animals was expected.' The number of degrees of freedom is then taken as two less than the number of groups after this combination. The test might seriously exaggerate the apparent significance of χ^2, since that statistic is not recalculated from the combined groups. Only a full amalgamation of groups and recalculation of χ^2 can give an unbiased test. In a short series of tests the data will generally be insufficient to show any significant departure from the fitted line; if χ^2 calculated according to equation (4·2) is well below the significance level for $(k-2)$ degrees of freedom (Table VI), no further test need be made and variances of estimates may be used without any heterogeneity factor. When a large value of χ^2 occurs, it may be necessary to recalculate on grouped data, or at least to investigate the possibility that an exaggerated effect of one or two anomalous observations is responsible.

No example of the computations for the fitting of the regression line is given here, as the initial determination of a provisional line is the only stage likely to trouble those who are familiar with the standard technique. A good illustration has been given by Bliss (1938) in the analysis of data on the toxicity of sodium fluoride to grasshoppers; the modified χ^2 test which he advocates has been criticized in the preceding paragraph, but inspection of these data shows there to be no significant heterogeneity. The approximate fiducial limits assigned to the LD 50 in this example are too narrow on account of the high variance of b (in fact $g = 0.44$); the exact formula (4.6) should have been used, and this gives limits of 0.062 and 0.135 mg./g. instead of 0.070 and 0.125 mg./g. Bliss has analysed these data both as individual records and by grouping in various ways; the estimated values of b differ quite widely, though they always lie within the range of sampling variation, but the estimates for the LD 50 are remarkably consistent.

A common practice in tests of this nature is to express the doses as amounts per unit weight of the test subject, in order to make some allowance for the varying size of these and their consequent probable variation in resistance. Thus the doses of sodium fluoride in the example used by Bliss were used as mg./g. of body weight. As an approximate method of adjustment this is not unreasonable, but a more exact approach would be to use the logarithm of the actual rather than the proportional dose as the dosage measure, and to make use of body weight as a concomitant variate. The assumption that resistance to the poison is directly proportional to body weight is thereby avoided, and instead the influence of body weight is estimated from the data. The computations may be carried out as in § 31, so that a probit plane is determined which relates mortality to log dose and log body weight; if desired, the effect of body weight may then be averaged out (§ 44).

In the light of the above discussion, little difficulty should be encountered in applying the methods of earlier chapters to individual mortality records. The various statistical techniques and formulae appropriate to comparisons of poisons, the action of mixtures, and multifactorial data can all be adopted even when the number of test subjects per batch is reduced to one. Obtaining a set of provisional probits for starting the computations is more troublesome when two or more dosage factors have to be considered in the same analysis, but instances of this are not likely to be encountered except by the more experienced workers to whom the estimation of some reasonable values will not be an insuperable obstacle; useful as it is to make a good first approximation, a poor one only delays the obtaining of satisfactory estimates, by requiring additional cycles of computations, and does not invalidate the final analysis.

44. THE AVERAGE KILL

Though not usually of great interest in insecticidal and fungicidal studies, the expected kill when a selected average dose is given, but the amounts received by individual test subjects vary, is sometimes required in other applications of the probit method. If the population has a true median lethal dose whose logarithm is μ, and the tolerance values of the log dose are normally distributed about this with variance σ^2, the true probit regression equation is

$$Y = 5 + (x - \mu)/\sigma,$$

more commonly written as

$$Y = \alpha + \beta x,$$

where $\beta = 1/\sigma$. Suppose now that the logarithms of the doses received by the subjects are normally distributed about a mean ξ with variance γ^2. Then the proportion killed will be the proportion receiving doses greater than their tolerance values. It might at first be thought that the result would be obtained by substituting ξ for x in the regression equation, but further consideration shows that the proportion also depends upon γ, being nearer to 50 % when γ is large compared with σ. By integration, the proportion is found to be that whose probit is $\overline{Y}$, where

$$\overline{Y} = 5 + (\alpha - 5 + \beta\xi)/(1 + \beta^2\gamma^2)^{\frac{1}{2}}. \tag{9.1}$$

In practice α and β have to be replaced by estimates a and b derived from experimental data. For example, if routine probit analysis has given a line

$$Y = 3 + 2x,$$

corresponding to a $\log \mathrm{LD} 50$ of 1 and a variance of $\frac{1}{4}$ for the distribution of the logarithms of individual tolerances, equation (9·1) gives

$$\overline{Y} = 5 + (2\xi - 2)/(1 + 4\gamma^2)^{\frac{1}{2}}.$$

Hence, if individuals are given doses whose logarithms are normally distributed about a mean of 1, the expected proportion killed is 50 % ($\overline{Y} = 5$), whatever the variance of the dosage distribution may be. If the mean log dose given is 2, the probit of the expected kill is nearly 7 (97·7 %) when γ, the standard deviation of the distribution of log doses, is very small, decreases to 6·41 (92·1 %) when γ is $\frac{1}{2}$, and decreases still further to a limiting value of 5 as γ increases indefinitely.

A slightly more complex situation arises when the mortality probit has been expressed, as in § 31, in terms of two dosage factors, say by the equation

$$Y = a + b_1 x_1 + b_2 x_2.$$

If the dosages to which the population is exposed are such that x_1 can be controlled but x_2 is normally distributed about ξ_2 with variance γ_2^2, the mortality probit is still linearly related to x_1, but the regression coefficient is reduced to

$$b_1' = b_1/(1 + b_2^2 \gamma_2^2)^{\frac{1}{2}}.$$

The one-factor regression equation is then

$$\overline{Y} = 5 + (a - 5 + b_2 \xi_2)/(1 + b_2^2 \gamma_2^2)^{\frac{1}{2}} + b_1' x_1. \qquad (9·2)$$

This last equation is useful when x_2, though formally a 'dose factor', is in fact a measurable characteristic of the individuals tested. For example, the potency of a poison may depend not only on the concentration but also on the weight of the subject. By subdivision of the data into weight classes, or by using individual records as described in § 43, the weight (or perhaps the logarithm of the weight) can be introduced as an x_2 into the probit regression equation. In order to predict the effect of any chosen concentration on a random selection of subjects, the contribution of weight to the equation must be averaged by means of equation (9·2). For this purpose ξ_2 and γ_2 may be estimated from the animals used for the toxicity test, provided that these were randomly selected from the whole population, or alternatively the parameters of the weight distribution may be estimated from measurements of a subsidiary random sample on which no toxicity tests have been made.

45. THE PARKER-RHODES EQUATION

A. F. Parker-Rhodes has made an extensive series of tests of the toxicity of metallic salts and other related compounds to spores of *Macrosporium sarcinaeforme* and *Botrytis allii*, and also to *Bacillus agri*. As a result of these tests, he has formulated a 'Theory of Variability', intended to explain, at least in part, the comparative toxic effects of different compounds (especially salts of the same metal) in terms of their chemical constitutions. Details of his ingenious theory, with experimental results, have been presented by the author in a series of papers (1941, 1942*a*, *b*, *c*, 1943*a*, *b*) to which the reader must be referred for the chemical and biological arguments. Discussion of the validity of these arguments is outside the scope of this book, but a brief outline of the statistical implications may be helpful to those concerned with fungicidal investigations of a like nature, the more particularly as Parker-Rhodes's treatment of this aspect seems scarcely adequate. For this purpose his notation will be brought into line with that of earlier chapters.

It has so far usually been assumed that the logarithm of λ, the tolerance of an individual test organism, is normally distributed, though mention has been made of the case of λ itself being normally distributed. O'Kane *et al.* (1930, 1934) published data from insecticidal studies which suggested that the logarithm of the mortality probit, rather that the probit itself, was linearly related to the log concentration. This theory may be expressed by the statement that the probit is proportional to a power of the dose. Parker-Rhodes's theory introduces the generalization that λ^i, a fractional power of the individual tolerances, shall be normally distributed. The relationship between mortality probit and dose thus becomes

$$Y = \alpha + \beta\lambda^i, \tag{9·3}$$

of which O'Kane's equation is a particular case (Bliss, 1935c). Parker-Rhodes defines i (for which he uses the symbol α) as the *index of variation*; $i = 0$ is a limiting case corresponding to the normal distribution of log tolerances, and $i = 1$ gives a normal distribution of tolerances. In general, the normalizing transformation for the dosage is

$$x = \lambda^i,$$

the tolerances on the x-scale being normally distributed about a mean value μ with variance $\sigma^2 = 1/\beta^2$. This normal distribution can only be an approximation and must break down for small doses (assuming i to be positive); λ^i cannot be less than zero, and equation (9·3) therefore suggests that the kill approaches a minimal value represented by $Y = \alpha$ as λ tends to zero.

He defines the *variability** of the spores relative to the toxic substance under investigation as

$$W_i(\lambda) = \sigma^2/i^2\mu^2, \tag{9·4}$$

which for the limiting case of $i = 0$ takes the form

$$W_0(\lambda) = \sigma^2. \tag{9·5}$$

Using estimates of σ and the median lethal dose obtained from the observations, an estimate of the variability is

$$U_i(\lambda) = 1/i^2b^2m^2, \tag{9·6}$$

or, in the limiting case,

$$U_0(\lambda) = 1/b^2. \tag{9·7}$$

The properties of the variability and index of variation in relation to the chemical and fungicidal behaviour of the

* This definition of variability is likely to be confused with other aspects of the variation in the behaviour of the spores, and the name is an unfortunate choice for a precisely defined quantity. In order to follow Parker-Rhodes, the word will be used in the strict sense for the remainder of this section.

compounds under investigation have been developed in the
second of the series of papers (Parker-Rhodes, 1942*a*). The
two chief results are:

'It is shown that the variability of a given population of spores
to a compound which can penetrate the spore wall is less than to
any other compound of the same element which cannot do so
unless it undergoes one or more reactions on the surface of the
spore with substances secreted by it, and that the greater the
number of such successive reactions that are required to bring
it to a permeable form, the greater will the variability be.

'It is shown that the variability of the spores to any compound
is proportional to the square of the number of atoms of the
effective element in a molecule of that compound, and that
the index of variation is inversely proportional to that number,
provided only one compound is permeative.'

Any study of a series of compounds in relation to the Theory
of Variability must therefore pay special attention to the indices
of variation and to the variabilities. In the most general case,
not only have the parameters α and β to be estimated from the
data, but also the index of variation. Experience indicates,
however, that the index takes either simple fractional values or
values so near these as to be indistinguishable from them. Indeed,
Parker-Rhodes states (1943*a*): 'On *a priori* grounds, however,
it appears that the index of variation must always be a rational
number, and is more likely to be a simple fraction than a complicated one.' For example, in addition to the common values
of 0 and 1, he finds (1942*b*) values very close to $\frac{1}{4}$ and $\frac{1}{2}$ (but see
Ex. 32 below). Negative values, $i = -\frac{1}{3}$ and $i = -\frac{2}{3}$, have also
been found (1943*a*), in association with negative values of β.
A negative index of variation implies that, as the concentration
is increased, the kill will approach, asymptotically, a maximum
value corresponding to $Y = \alpha$; if this level is less than 50 %,
formal solution of the equations will give an imaginary value
for the median lethal concentration, the value of m (measured on
the $x = \lambda^i$ scale) being negative. There may in truth sometimes
be such a limiting mortality, but any expression of the situation
in terms of 'imaginary LD 50's' is useless. In practice it seems

likely that, as in Parker-Rhodes's examples, at the higher concentrations some other phase of toxic action will supervene. Parker-Rhodes's statement that 'An imaginary value may be taken to imply absence of fungicidal potency' is misleading, even if the word 'potency' is considered to refer only to the one phase of toxic action, for though this phase may fail to attain an LD 50 it may well possess, say, an LD 45.

If the view that the index of variation is a simple rational fraction is accepted, the necessity of forming a statistical estimate from the data may be avoided; the subsequent calculations are then much simplified and the precision of estimation of α and β is increased. If the value of i can be decided either from previous experience or from a preliminary inspection of the data, the technique for estimating α and β is exactly as described in § 17, using $x = \lambda^i$.

Parker-Rhodes has apparently based his formula (1942b, c) for testing the significance of differences between two estimates of variability on the belief that, when there is no heterogeneity, for $i = 0$

$$V(U^{-\frac{1}{2}}) = 1/S_{xx}, \tag{9.8}$$

and for other values of i

$$V(U^{-\frac{1}{2}}) = i^2 m^2/S_{xx},$$

m being, as usual, defined by

$$m = \bar{x} + (5 - \bar{y})/b.$$

The first of these equations is correct, but the second is easily seen to be entirely wrong. From equation (9.6)

$$U^{-\frac{1}{2}} = ibm$$
$$= i(5 - \bar{y} + b\bar{x}),$$

whence
$$V(U^{-\frac{1}{2}}) = i^2 \left(\frac{1}{Snw} + \frac{\bar{x}^2}{S_{xx}} \right). \tag{9.9}$$

The criterion of significance at the 5 % level given by Parker-Rhodes (1942b, p. 141) should therefore be amended to read

$$\frac{U_1 + U_2 - 2(U_1 U_2)^{\frac{1}{2}}}{U_1 U_2 \{V(U_1^{-\frac{1}{2}}) + V(U_2^{-\frac{1}{2}})\}} \geqslant 3.84,$$

where the two variances are calculated according to equations (9·8) or (9·9), whichever is appropriate; the test may alternatively be considered as a test of the difference between $U_1^{-\frac{1}{2}}$ and $U_2^{-\frac{1}{2}}$ by means of their standard errors. When the heterogeneity χ^2 is significant, all variances should be multiplied by the heterogeneity factor, and the tests based on normal distribution of errors should be changed to the corresponding t-tests. In the early papers of the series, Parker-Rhodes used a conventional value of 50 or 100 for the heterogeneity factor, but in a note to the fourth (1942 c), he recognized this to be wrong and suggested an amendment.*

Ex. 31. *Variability of* Macrosporium sarcinaeforme *to hydrogen sulphide, sodium dithionite, and sodium tetrathionate.* As part of the data from an investigation into the toxicity of various sulphur compounds to $M.$ *sarcinaeforme,* Parker-Rhodes has published results for series of concentrations of hydrogen sulphide, sodium dithionite, and sodium tetrathionate (1942 b, Tables 2, 5 and 6). The concentrations there given are expressed in arbitrary units, since the only use made of the data is for the estimation of variability, a quantity independent of the unit of concentration. Parker-Rhodes uses x instead of λ to represent concentration, q instead of p to represent proportionate mortality, and his n' is the number of sets of 50 spores counted. In his Table 8, Parker-Rhodes gives 0·95, 0·48 and 0·24 for the estimated indices of variation for the three compounds; these values are estimated from the data, by a method which is discussed and criticized below, but, in view of what has been said about likely values for the index of variation, they may reasonably be taken as 1, $\frac{1}{2}$ and $\frac{1}{4}$. Table 29 gives details of the calculations for estimating the variability for sodium tetrathionate, on the assumption that $i = \frac{1}{4}$. The first stages are exactly as in the ordinary probit

* In the next paper (1943 a) Parker-Rhodes introduced a form of the test equivalent to the use of the heterogeneity factor, though still involving the incorrect form for the variance of $U^{-\frac{1}{2}}$. The reference to the present writer at this point unintentionally suggests that he advocated the adoption of an 'admittedly false assumption'; in fact, his advice was not so revolutionary but only recommended bringing the test into line with other uses of the heterogeneity factor.

computations, except that x is taken as $\lambda^{\frac{1}{4}}$ instead of $\log\lambda$.*
Apparently Parker-Rhodes adjusted the values of p, the pro-
portion of spores failing to germinate, so as to take account of
a control mortality, but he gives no details, and for this example
the modifications introduced in Chapter 6 have been ignored.

TABLE 29. Estimation of Variability of *Macrosporium*
sarcinaeforme relative to Sodium Tetrathionate

$x = \lambda^{\frac{1}{4}}$	n	p	Empirical probit	Y	nw	y	nwx	nwy
1·00	400	4·8	3·34	3·4	95	3·34	95·00	317·30
1·50	500	11·5	3·80	4·0	219	3·82	328·50	836·58
2·00	500	43·6	4·84	4·7	308	4·84	616·00	1490·72
2·51	500	64·1	5·36	5·3	308	5·36	773·08	1650·88
2·99	500	80·4	5·86	5·9	236	5·86	705·64	1382·96
3·50	500	91·5	6·37	6·5	135	6·36	472·50	858·60
					1301		2990·72	6537·04

$1/Snw = 0{\cdot}0007686395, \quad \bar{x} = 2{\cdot}29879, \quad \bar{y} = 5{\cdot}02463.$

$Snwx^2$	$Snwxy$	$Snwy^2$
7523·79	15837·47	33884·16
6875·02	15027·25	32846·19
648·77	810·22	1037·97
		1011·85

$$26{\cdot}12 = \chi^2_{[4]}$$

$$b = 1{\cdot}24886,$$
$$Y = 2{\cdot}1538 + 1{\cdot}2489x,$$
$$m = 2{\cdot}279,$$
$$\frac{1}{Snw} + \frac{\bar{x}^2}{S_{xx}} = 0{\cdot}000769 + 0{\cdot}008145$$
$$= 0{\cdot}008914.$$

The computations present no new features until the regression
coefficient, b, and the median lethal x-value, m, have been calculated. The fitted equation is

$$Y = 2{\cdot}1538 + 1{\cdot}2489\lambda^{0{\cdot}25}. \qquad (9{\cdot}10)$$

It follows that $\quad U^{-\frac{1}{2}} = ibm = 0{\cdot}25 \times 1{\cdot}2489 \times 2{\cdot}279$
$$= 0{\cdot}712,$$
whence $\quad U_{\frac{1}{4}}(\lambda) = 1{\cdot}97$

is the estimated variability for sodium tetrathionate. Parker-
Rhodes's estimate is 1·92; the difference is presumably due to

* x is easily obtained as the antilogarithm of $\frac{1}{4}\log\lambda$.

different degrees of approximation in the maximum likelihood estimation of the regression line and to the slightly different indices of variation used, 0·25 and 0·24.

For the other two compounds, $U^{-\frac{1}{2}}$ and U have been similarly calculated, and their values are shown in Table 30. In each case there is evidence of heterogeneity, the values of χ^2 being, for hydrogen sulphide

$$\chi^2_{[3]} = 8\cdot60,$$

TABLE 30. Variability of *Macrosporium sarcinaeforme* relative to Hydrogen Sulphide, Sodium Dithionite, and Sodium Tetrathionate

Compound tested	Index of variation	$V(U^{-\frac{1}{2}})$	$U^{-\frac{1}{2}}$	Variability (U)	
				Present calculations	Parker-Rhodes
Hydrogen sulphide	1	0·0548	3·114 ± 0·234	0·103	0·104
Sodium dithionite	$\frac{1}{2}$	0·0122	1·895 ± 0·110	0·278	0·272
Sodium tetrathionate	$\frac{1}{4}$	0·00352	0·712 ± 0·059	1·97	1·92

for sodium dithionite (one concentration of the twelve tested was so high as to give zero weight)

$$\chi^2_{[9]} = 66\cdot44,$$

and for sodium tetrathionate

$$\chi^2_{[4]} = 26\cdot12.$$

Mean squares derived from these do not differ significantly, and the heterogeneity factor, obtained from

$$\chi^2_{[16]} = 101\cdot16,$$

has the value 6·32. Now from Table 29

$$\frac{1}{Snw} + \frac{\bar{x}^2}{S_{xx}} = 0\cdot008914$$

for sodium tetrathionate, whence, by equations (9·9),

$$V(U^{-\frac{1}{2}}) = 6\cdot32 \times (0\cdot25)^2 \times 0\cdot008914$$
$$= 0\cdot00352.$$

The standard errors of $U^{-\frac{1}{2}}$ make it clear that variability differences between the three compounds are highly significant. The variance of the difference between any pair of values of $U^{-\frac{1}{2}}$ is the sum of the corresponding variances tabulated in the third column; if an exact test were needed, a t-test with 16 degrees of freedom (the number of degrees of freedom on which the heterogeneity factor is based) would show whether or not the difference was significantly greater than zero.

When the estimation of the index of variation from the data seems necessary, a more complex procedure must be adopted. Parker-Rhodes (1942a) has discussed this problem and has suggested a method of obtaining the maximum likelihood estimate. Comparison with the discussion of the general maximum likelihood equations in Appendix II shows his method to be at fault in several respects. The estimates obtained by it will frequently not differ greatly from the true maximum likelihood values, but since the correct method is no more difficult or laborious to apply there seems every reason for preferring it.

Equations (II, 3) may be adapted to the estimation of the parameters of equation (9·3) in much the same way as they were adapted in § 28 to give equations (6·4). The most convenient method of computation is first to tabulate an additional variate*

$$x' = x \log_e \lambda.$$

If a provisional value is taken for i, and a provisional regression line of probits against $x = \lambda^i$ plotted, weights, nw, and working probits, y, can be derived for each concentration by the usual process. The equations

$$\left. \begin{array}{lll} a'Snw & +b\delta iSnwx' & \doteq Snwy, \\ bSnw(x-\bar{x})^2 & +b\delta iSnwx'(x-\bar{x}) = Snw(y-\bar{y})\,(x-\bar{x}), \\ a'Snwx' + bSnwx'(x-\bar{x}) + b\delta iSnwx'^2 & = Snwyx', \end{array} \right\}$$

$$(9·11)$$

* Note that either natural logarithms must be used in the separate values of x', or, if logarithms to base 10 are used, the sums of squares and products in equations (9·11) and all other derived quantities must be adjusted by multiplying by 2·30259 or 5·30190 according to whether they involve x' or x'^2.

may then be solved for a', b and $b\delta i$; δi is an adjustment to i giving a revised approximation to the maximum likelihood estimate. If i is written for $i+\delta i$, the new approximation to the probit regression equation is

$$Y = a' + b(x-\bar{x}) = a' + b(\lambda^i - \overline{\lambda^i}).$$

The process may be repeated with the new i and the new regression equation as provisional estimates; further cycles may be computed until δi becomes negligible and successive values of a' or b are not appreciably different. In the limit a' is equal to the mean probit, $\bar{y}$, but this is not true at intermediate stages. Careful choice of the provisional i with which to start should ensure that at most three cycles, and often only one or two, are needed.

Estimation of the variances of the index of variation and the variability will not be discussed in detail here. A matrix of variances and covariances for the three parameters a', b and i can be obtained by the methods of Appendix II. Unfortunately, for most experimental data discrimination between the effects of small alterations in i and small alterations in a' and b in the neighbourhood of the maximum likelihood estimates is very difficult; in other words, there is usually a close correlation between the sampling variations of the parameters. The variances of a and b are therefore very much greater than they would have been if i were known a priori, and the variance of i is often too great for any reliable estimate to be obtained. Without data vastly more extensive than are usually available, any precise determination of the index of variation will seldom be possible. On the other hand, the variability appears to be a more easily determinable quantity and not very sensitive to small changes in i, since the corresponding changes in the other parameters will often compensate.

Ex. 32. *Maximum likelihood estimation of index of variation of* Macrosporium sarcinaeforme *relative to sodium tetrathionate.* Further examination of the sodium tetrathionate results analysed in Ex. 31 indicates that the value of $\frac{1}{4}$ taken for the index of variation is substantially higher than the maximum likelihood estimate. Three cycles of computations for equations (9·11) have been carried out, the last starting from a provisional $i = 0·1$; these have given the revised estimates

$$a' = 5·0478, \quad b = 4·8304, \quad i = 0·1017,$$

whence the probit equation is

$$Y = -1·6706 + 4·8304\lambda^{0·1017}. \tag{9·12}$$

As a measure of heterogeneity,

$$\chi^2_{[3]} = 13·59$$

is obtained, giving 4·53 as the heterogeneity factor. The significance of the difference between the estimated index of variation and its previously assumed value may be tested by comparison of χ^2 in the two analyses. The test is made in the form of a variance ratio test, as shown in Table 31; the difference

TABLE 31. Test of Significance for the Difference between the Estimated Index of Variation and its Previously Assumed Value

Residual variation	D.F.	Sum of squares	Mean square
Removed by i	1	12·53	12·53
After fitting a, b, i	3	13·59	4·53
After fitting a, b	4	26·12	

between the heterogeneity χ^2 values in Exs. 31 and 32 measures the improvement due to fitting i (1 degree of freedom) and should be compared with the residual mean square, 4·53. Reference to Fisher and Yates (1943, Table V) shows the ratio of mean squares to be non-significant, but the test is not very sensitive since so few degrees of freedom are available, and there is at least some indication of an improvement. In Table 32 are compared the observed percentage mortalities with estimates from equations (9·10) and (9·12); the latter is seen to give the better agreement with observation.

The matrix of variances and covariances for the three parameters is

$$V = \begin{pmatrix} 4·13 & 11·35 & -0·182 \\ 11·35 & 31·36 & -0·500 \\ -0·182 & -0·500 & 0·0080 \end{pmatrix}$$

whence the standard deviations, each based on only three degrees of freedom, are $\pm 2\cdot03$, $\pm 5\cdot60$ and $\pm 0\cdot089$, values too large to be of any practical use. Though $i = 0\cdot25$ is very different from $i = 0\cdot1017$, the standard deviation of the latter estimate is so great as not to discriminate clearly between the two values, and the data cannot be said definitely to contradict the first assumption. From equation (9·12), $U^{-\frac{1}{2}}$ may be estimated, and,

TABLE 32.   Comparison of Two Probit Equations fitted to Sodium Tetrathionate-*Macrosporium sarcinaeforme* Tests

(1) $Y = \quad 2\cdot1538 + 1\cdot2489\lambda^{0\cdot25}$
(2) $Y = -1\cdot6706 + 4\cdot8304\lambda^{0\cdot1017}$

λ	n	Percentage Mortality		
		Observed	Expected (1)	Expected (2)
1	400	4·8	5·5	3·3
5	500	11·5	16·4	16·3
16	500	43·6	36·4	39·5
40	500	64·1	61·6	64·0
80	500	80·4	81·3	80·8
150	500	91·5	93·6	91·5

by formulae not shown here, the matrix V may be used to give an estimate of the variance of $U^{-\frac{1}{2}}$. The result of these operations is

$$U^{-\frac{1}{2}} = 0\cdot678 \pm 0\cdot050,$$

which is in good agreement with the estimate made in Ex. 31, thus illustrating the stability of the variability for different values of i. The reduction in the standard error of $U^{-\frac{1}{2}}$ is entirely due to the smaller heterogeneity factor (4·53 instead of 6·32), and no doubt a more reliable standard error could be obtained by using a composite heterogeneity factor from several analyses so as to have a greater number of degrees of freedom. The variability as now estimated is 2·18 instead of 1·97.

Chapter 10

GRADED RESPONSES

46. The Linear Dosage-Response Curve

THE discussion in earlier chapters has been concerned only with quantal characteristic responses. In many biological investigations it is possible not merely to state that a subject has responded to the treatment applied but also to measure the magnitude of that response. Vitamin preparations, for example, are often assayed by comparing the weight increases of rats fed for a specified period on suitable doses with the weight increases of other rats fed for the same time on a similar range of doses of a standard preparation with known vitamin content. Again, insulin may be assayed in terms of the percentage fall in blood sugar of injected rabbits. All such data could be reduced to a quantal form by a simple dichotomy of the measurements into those greater than and those not greater than an arbitrarily selected value; the relationship between the dose and the percentage of subjects whose responses exceed this value could then be studied by the methods of probit analysis already described. This procedure, however, would be very wasteful, as it discards entirely the information provided by the distribution of the magnitudes of responses within the 'greater than' and 'not greater than' classes.

When quantitative responses to a stimulus are measured over a sufficiently wide range of doses, some curvature of the relationship between dose and mean response will almost certainly become evident. Indeed, the response curve will frequently be sigmoid in type. Nevertheless, especially if the responses are plotted against a logarithmic dosage scale, the curve often does not differ appreciably from a straight line over a considerable section of the dosage range. The central section is usually that of chief interest to the experimenter, and he may reject the results for the more extreme levels of dose which give responses outside the linear range in order to have data which can be analysed by the ordinary linear regression technique (Fisher, 1944, §§ 25, 26). The accurate estimation of the regression slope requires that the whole of the linear section should be included in the tests, but pre-existing knowledge may enable economies in experimental material to be effected by omitting very extreme doses which are almost certain to lie outside this range.

Extreme responses will thus be avoided, and it may then be reasonable to assume that the variance of the response of each subject is constant, irrespective of the dose. A linear regression equation of mean response on dosage may therefore be calculated, weighting the mean response at each dosage in proportion to the number of subjects. Since these weights are independent of the expected responses, the complications of probit analysis do not arise and no method of successive approximations is required in order to obtain the maximum likelihood solution. Burn (1937) and Coward (1938) have used regression analysis extensively in a variety of problems of biological assay from quantitative responses. Bliss and Marks (1939 a, b) have given an excellent description of the statistical analysis of an insulin assay, showing details of the computations and discussing many important points, including the use of covariance analysis for improving the precision of the estimate of relative potency by making adjustments for preliminary variation in relevant measurable characteristics of the experimental rabbits. The same considerations in respect of fiducial limits as are outlined for quantal response data in § 19 arise for quantitative responses also, and the general formula (4·7) may be applied (Irwin, 1943; Fieller, 1944).

Details of the application of standard linear regression technique to assay data will not be given here, for, though the underlying theory is comparatively simple, the methods could not be adequately illustrated without many examples, and a full account falls outside the scope of this book. In addition to references already given, Bliss (1940 a), Bliss and Rose (1940), Fieller (1940) Finney (1945), and Irwin (1937) may be consulted for useful examples of experimental planning and the statistical reduction of the results. Before leaving this topic, however, one small

point of interest may be noted in connexion with what has come
to be known as the *four-point assay*, the very simple type of
assay in which a test material and a standard are each used at
two rates only; the difference between the higher and lower rates
is the same for both materials on the dosage scale used, thus
implying that, if a logarithmic dosage scale has been adopted,
the two pairs of doses are in the same ratio. Gridgeman (1943)
and Wood (1944a) have shown that for this type of assay the
relative potency may be estimated even when there is an appreciable departure of the dosage-response relationship from
linearity over the range of doses tested. Provided that the
relationship can be adequately represented by a second-degree
equation, the formula for relative potency derived on an assumption of linearity still gives an unbiased estimate. The four-point
assay may nevertheless give misleading results because of the
very little evidence it supplies on the identity of shape of the response curves for the two materials tested; the conditions of its
usefulness and validity have been discussed by Finney (1944b),
Gridgeman (1944) and Wood (1944b).

47. QUANTITATIVE RESPONSES AND THE PROBIT TRANSFORMATION

When quantitative responses are studied over an unrestricted
range of doses, they will frequently be found to show a sigmoid
relationship with dose, and by a suitable choice of dosage scale
(again often logarithmic) the relationship may usually be made
to approximate to the normal sigmoid form. Though the probit
transformation was devised as an aid to the analysis of quantal
response data, its property of converting a normal sigmoid curve
into a straight line is not dependent upon the source or nature
of the data, and the same end will therefore be achieved if the
sigmoid represents the relationship between dosage and a quantitative response. There is, however, an important difference
between the statistical procedures for the two types of data.
For quantal responses the proportion of subjects responding
to any dose is assessed as the ratio of the number showing the
characteristic response to the total number receiving the dose;

for quantitative responses the mean response to each dose can be calculated from the data, but in general no maximum possible response is known and consequently no proportional response can be directly calculated. Sometimes, as in Ex. 33 below, one subject or set of subjects can be given a dose such that each will show its individual maximum, but this can only be an estimate of the mean for all subjects, since it is itself subject to the natural variability of the population. The difficulty is similar to that of correcting quantal data for the proportion of responses amongst the controls (Chapter 6), but has to be faced relatively more frequently. Indeed, the rarity of investigations in which quantitative responses can be reckoned as proportions of a known maximum probably accounts for the failure to realize the possibility of applying the probit transformation to quantitative data (Bliss, 1941; Finney, 1943 b).

The variance of the response of all test subjects receiving the same dose can usually be estimated empirically from the separate observed responses to that dose, and thus independently of the expected response to that dose; the variance of the mean response then takes the form v/n, where v is the variance of a single response and n is the number of responses contributing to the mean. At low dosages the mean response tends to zero, and, unless negative responses of individual test subjects are possible, v must also become very small; at high dosages also, when the mean response approaches its maximum, v may again become small. Nevertheless, unless very extreme doses are used, v may often be taken as substantially constant and may then be assessed by pooling estimates for different dose levels.*

If the mean response to n subjects tested at dosage x (on the normalizing scale) is u, the expected value of u may be written

$$U = HP; \qquad (10\cdot1)$$

---

* An approximate test of the heterogeneity of a set of variance estimates $v_1, v_2, ..., v_k$, with degrees of freedom $f_1, f_2, ..., f_k$ is given by treating

$$2\cdot3026(f \log v - Sf_i \log v_i)$$

as a χ^2 with $(k-1)$ degrees of freedom. The summation is to be taken over the k values, and $f = Sf_i$, $fv = Sf_i v_i$; v is the pooled estimate of variance to be used if the test discloses no heterogeneity (Bartlett, 1937).

here H is the mean limiting response for high dosages, and P, the proportionate response at this dosage, is as defined by equation (3·1). When H is known, and therefore does not have to be estimated from the data,

$$p = u/H \qquad (10\cdot2)$$

is an estimate of P. The probits of the several values of p may then be fitted by a linear regression on x by the method of § 17, except that the weight to be attached to each point is not nZ^2/PQ but nH^2Z^2/v. The standard errors of parameters estimated from the analysis are proportional to $\sqrt{v}$ and are based on the same number of degrees of freedom as v instead of pertaining to a normal distribution. Instead of a χ^2 test of the significance of the departures of the observations from the fitted probit regression line, a variance ratio test must be used, as illustrated in Ex. 33.

In the present chapter, a weighting coefficient, w, defined by

$$w = Z^2, \qquad (10\cdot3)$$

will be used instead of that defined by equation (3·4). Values of both Z and Z^2 are given in Table V for values of Y at intervals of 0·1 from 1·0 to 9·0; both functions are symmetrical about $Y = 5\cdot0$. For the simple problem of fitting the probit regression line when H is known, the only differences from the procedure of § 17 are the changed weighting coefficient and the introduction of a factor H^2/v into all weights, or v/H^2 into all variances, at the end of the analysis. The working probit is not affected by the change in weighting.

When H is unknown, there are three parameters to be estimated from the data, and the maximum likelihood equations for these may be derived and calculated in a form rather similar to equations (6·4). The equations are:

$$\left.\begin{array}{l}
\delta H . \left\{ n_h + Snw\dfrac{P^2}{Z^2} \right\} + a'H\,Snw\dfrac{P}{Z} + bH\,Snw(x-\bar{x})\dfrac{P}{Z} = n_h(h-H) + H\,Snwy\dfrac{P}{Z}, \\[3mm]
\delta H . H\,Snw\dfrac{P}{Z} \qquad\;\; + a'H^2Snw \qquad\qquad\qquad\qquad = H^2Snwy, \\[3mm]
\delta H . H\,Snw(x-\bar{x})\dfrac{P}{Z} \qquad\qquad + bH^2Snw(x-\bar{x})^2 = H^2Snw(x-\bar{x})\,(y-\bar{y}),
\end{array}\right\}$$

$$(10\cdot4)$$

where δH is an adjustment to a provisional value of H, and a', b are the estimates of the parameters of the probit regression line; h is the value of the mean response, u, for n_h test subjects treated so as to give the maximal response, and is thus itself an estimate of H when such 'controls' are available. The column headed Q/Z in Table II may be used to give P/Z by reading the entry corresponding to $(10 - Y)$ instead of to Y.

The inverse matrix of the coefficients in equations (10·4), when multiplied by v, gives the variances and covariances of the estimates of the three parameters H, a' and b. In general, a' is not the weighted mean probit, $\bar{y}$, but if the provisional value of H is near to the maximum likelihood estimate, or if the cycle of computations is repeated sufficiently often, δH becomes very small, a' becomes equal to $\bar{y}$, and b becomes the regression coefficient of y on x. An example will make the method clear.

Ex. 33. *The repellent effect of lime sulphur on the honeybee.* Butler *et al.* (1943) have discussed a series of experiments on the attractiveness or repellency to the honeybee of various constituents of orchard sprays. In one of these experiments individual cells of dry brood comb were filled with measured volumes of emulsions of lime sulphur in $M/1$ sucrose solution, and were placed in an experimental chamber. Seven different concentrations of lime sulphur were used, ranging from 1/100 to 1/1,562,500 by successive factors of one-fifth. Eight cells of each of these, together with eight cells of $M/1$ sucrose solution alone, were arranged in eight rows of eight to form an 8×8 Latin square, thus eliminating as far as possible any positional effects within the chamber. Additional non-experimental cells of sucrose solution were placed between the rows and around the square so that there should be no shortage of the unadulterated solution. About 100 honeybees were then released into the chamber for two hours, after which the volume of liquid within each cell was again measured. The difference between the two volumes, corrected by a very small amount representing the evaporation during the experiment, is the quantity taken up by the bees. Table 33 shows the arrangement of the experimental cells and the corrected uptake figures for each; totals of rows and columns of eight cells are also shown. The effectiveness of lime sulphur in repelling the honeybees is measured by comparing the total uptakes from the eight cells of each treatment, shown in Table 34.

TABLE 33. Lay-out and Results of Experiment on Uptake of Lime Sulphur by the Honeybee

(Uptake recorded in mg. per cell)

								Row totals
D 57	C 84	F 87	H 130	E 43	A 12	B 8	G 80	501
E 95	B 6	H 72	A 4	D 28	C 29	G 72	F 114	420
B 8	H 127	A 5	E 114	G 60	F 44	C 13	D 39	410
H 69	D 36	E· 39	C 9	A 5	G 77	F 57	B 14	306
G 92	E 51	D 22	F 20	C 17	B 4	A 4	H 86	296
F 90	A 2	C 16	G 24	B 7	D 27	H 81	E 55	302
C 15	F 69	G 72	B 10	H 81	E 47	D 20	A 3	317
A 2	G 71	B 4	D 51	F 71	H 76	E 61	C 19	355
Column totals 428	446	317	362	312	316	316	410	2907

The meaning of the symbols for the eight treatments is given in Table 34.

TABLE 34. Uptakes of Various Concentrations of Lime Sulphur by Honeybees

Treatment	Total of 8 cells (mg.)	Mean per cell (mg.)
A: $M/1$ sucrose + lime sulphur (1 %)	37	4·6
B: ,, ,, (0·2 %)	61	7·6
C: ,, ,, (0·04 %)	202	25·2
D: ,, ,, (0·008 %)	280	35·0
E: ,, ,, (0·0016 %)	505	63·1
F: ,, ,, (0·00032 %)	552	69·0
G: ,, ,, (0·000064 %)	548	68·5
H: $M/1$ sucrose alone	722	90·2
Standard error of mean		± 6·9

An analysis of variance of the 64 uptakes involves the assumption that the error variance is the same for all of them. In fact there are strong indications that the variance increases with decreasing concentration of lime sulphur, and especially that the uptakes of the four highest concentrations are much less variable than those of the four lowest. Experience has shown that the validity of conclusions from an analysis of variance is not seriously upset unless the differences in variability are very great, and special investigation of these data indicates that the final estimates obtained would not be much altered if they were based on a variance which itself depended upon the concentration. The standard procedure for the analysis of Latin square data (Fisher, 1944, § 49; Mather, 1943, § 29) has therefore been adopted as a means of estimating the variance per cell. Each row of eight cells in the experimental lay-out contains one cell of each treatment, as also does each column of eight cells, and the variation between the sets of row and column totals is eliminated in the statistical analysis.

The completed analysis of variance is given in Table 35. A total sum of squares of deviations of the 64 measurements about their mean is calculated as

$$57^2 + 84^2 + 87^2 + 130^2 + \ldots + 61^2 + 19^2 - (2907)^2/64 = 79730.$$

TABLE 35. Analysis of Variance of Data in Table 33

	D.F.	Sum of squares	Mean square
Rows	7	4,768	
Columns	7	2,808	
Treatments	7	56,160	8,023
Error	42	15,994	380·8
Total	63	79,730	

A component of this sum of squares representing the variation in row totals, and having 7 degrees of freedom, is then calculated from the totals shown in Table 33 as

$$(501^2 + 420^2 + \ldots + 355^2)/8 - (2907)^2/64 = 4768,$$

the divisor 8 occurring since each row has eight cells. Similar calculations give components for columns and for treatments, the latter being obtained from totals shown in Table 34 as

$$(37^2 + 61^2 + \ldots + 722^2)/8 - (2907)^2/64 = 56160.$$

The difference between these three components and the total is a sum of squares with 42 degrees of freedom which measures the residual variation, and the mean square obtained when this residual is divided by 42 is the estimate of error variance,

$$v = 380 \cdot 8.$$

A rapid inspection of the data is sufficient to show that the differences between the uptakes of the eight concentrations are much too large to be attributed to random variation. If a test of significance were required, it would consist in comparing the mean squares for 'Treatments' and 'Error'; according to Fisher and Yates's table of the 5 % levels of the variance ratio (1943, Table V), a ratio greater than $2 \cdot 24$ must be considered indicative of real differences between treatments, and in this analysis the ratio is over 20.

The standard error of the treatment means, each a mean for eight cells, is $\pm \sqrt{(v/8)}$, or $\pm 6 \cdot 9$ mg. per cell. The means are given in Table 34, and are also shown in Fig. 21 plotted against a logarithmic scale of dosage; on this scale, for convenience, the $0 \cdot 008 \%$ ($= 5^{-3} \%$) concentration has been taken as zero and other doses expressed relative to it by logarithms to base 5, so that

$$x = 3 + \log_5 \text{ (percentage concentration).} \qquad (10 \cdot 5)$$

Sucrose alone is a zero concentration of lime sulphur and therefore has an infinite negative value of x. Though a straight line would fit the points for the seven concentrations of lime sulphur tolerably well, a sigmoid of some type is necessary to show the eventual tapering away to zero uptake at very high concentrations and the approach to the uptake of unadulterated sucrose solution at very low concentrations. A straight line would be an entirely inadequate representation of the relationship much outside the range of concentrations tested, and indeed the points do suggest a sigmoid curve.

Apart from the reversal in the direction of the dose effect, so that increasing the dose decreases the response, the sigmoid required does not look to be markedly different in shape from the normal sigmoid illustrated in Fig. 4, and the frequent success of log dose as a normalizing transformation encourages its use here. In this experiment the test subjects of the earlier discussion are the cells, not the bees, so that there are eight 'subjects' at each concentration; the response, u, is the uptake of liquid from a cell during the course of the experiment. Since u decreases with increasing concentration from the value for sucrose at zero concentration to zero at very high concentrations, for the probit regression line b must be negative.

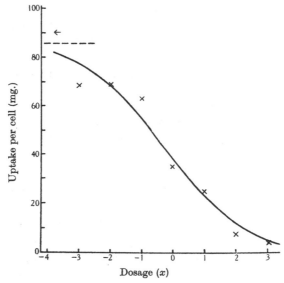

FIG. 21. Uptake of lime sulphur by the honeybee, showing normal sigmoid curve (Ex. 33). ← indicates uptake of zero concentration. Broken line indicates mean limiting response.

The mean uptake of liquid from the cells containing only sucrose solution is 90 mg. per cell, so that the estimate of the parameter H from these control cells alone is $h = 90$. On the other hand, the two lowest concentrations of lime sulphur tested gave responses of just under 70 mg. per cell, thus suggesting that H is not so great as 90. In obtaining equations (10·4) a provisional value of 85 was taken for H, a guess which proved to be remarkably good.

The computations then proceed as in Table 36. Values of u, the mean uptake at dosage x, are copied from Table 34, and $p = u/85$ is calculated for each. The empirical probits of p are tabulated, and are plotted in Fig. 22, in which figure a provisional regression line is drawn. From this line expected probits, Y, are read, and weights and working probits are determined in the usual manner, except that w must be taken from the Z^2 column of Table V instead of from Table II; thus for the third dosage $Y = 4\cdot4$, and the weight is therefore $nw = 8 \times 0\cdot111$. The quantities P/Z are obtained from Table II as the values of Q/Z which correspond to $(10 - Y)$.

TABLE 36. Computations on Uptake of Lime Sulphur by Honeybees

x	n	u	$p(H=85)$	Empirical probit	Y	nw	y	P/Z	nwx	nwy	nwP/Z
3	8	4·6	0·05	3·36	3·4	0·10	3·36	0·49	0·30	0·3360	0·0490
2	8	7·6	0·09	3·66	3·9	0·38	3·69	0·62	0·76	1·4022	0·2356
1	8	25·2	0·30	4·48	4·4	0·89	4·48	0·82	0·89	3·9872	0·7298
0	8	35·0	0·41	4·77	4·9	1·26	4·77	1·16	0·00	6·0102	1·4616
−1	8	63·1	0·74	5·64	5·4	1·08	5·63	1·78	−1·08	6·0804	1·9224
−2	8	69·0	0·81	5·88	5·9	0·57	5·88	3·07	−1·14	3·3516	1·7499
−3	8	68·5	0·81	5·88	6·4	0·18	5·67	6·14	−0·54	1·0206	1·1052
$-\infty$	8	90·25	—	—	—	—	—	—	—	—	—
						4·46			−0·81	22·1882	7·2535

$$\bar{x} = -0\cdot1816, \quad n_h(h-H) = 8 \times (90\cdot25 - 85) = 42\cdot00$$

$Snwx^2$	$Snwxy$	$SnwxP/Z$	$SnwyP/Z$	$SnwP^2/Z^2$
8·2900	−8·0458	−7·3898	38·6543	18·0440
0·1471	−4·0297	−1·3173		$8\cdot0000 = n_h$
8·1429	−4·0161	−6·0725		26·0440

The products nwx, nwy and nwP/Z are next calculated, entered in Table 36, and added. The various sums of squares and products shown in the last portion of Table 36 are found in the usual manner, so that, for example,

$$Snw(x - \bar{x})^2 = 8\cdot1429.$$

These quantities are then multiplied by 85 or 85^2 where necessary, and used to construct equations (10·4); in particular

$$n_h(h - H) + H\,Snwy\frac{P}{Z} = 8 \times 5\cdot25 + 85 \times 38\cdot6543$$
$$= 3327\cdot62.$$

The equations are

$$
\left.
\begin{aligned}
26 \cdot 0440 \delta H + \quad 616 \cdot 55 a' - \quad 516 \cdot 16 b &= \quad 3327 \cdot 62 \\
616 \cdot 55 \quad \delta H + 32224 \quad a' \qquad\qquad &= \quad 160310 \\
-516 \cdot 16 \quad \delta H \qquad\qquad\quad + 58832 \quad b &= -29016
\end{aligned}
\right\} \quad (10 \cdot 6)
$$

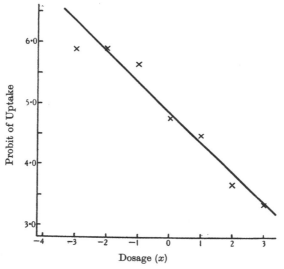

Fig. 22. Data of Fig. 21 transformed to probit scale, showing probit regression line (Ex. 33).

By the usual process, the inverse matrix of coefficients is found, and this, with a factor v, gives the variances and covariances of the estimate of the parameters in the form

$$
V = 380 \cdot 8 \times
\begin{pmatrix}
0 \cdot 1028925 & -0 \cdot 001968668 & 0 \cdot 000902723 \\
-0 \cdot 001968668 & 0 \cdot 00006869979 & -0 \cdot 00001727202 \\
0 \cdot 000902723 & -0 \cdot 00001727202 & 0 \cdot 00002491755
\end{pmatrix}. \quad (10 \cdot 7)
$$

The solutions of equations (10·6), obtained as the sum of products of each row of this matrix (omitting v) with the quantities on the right-hand side of the equations, are:

$$
\delta H = 0 \cdot 5966, \quad a' = 4 \cdot 9634, \quad b = -0 \cdot 4880.
$$

Hence the improved estimate of H is $H + \delta H = 85\cdot60$, and the probit regression line is estimated to be

$$Y = 4\cdot9634 - 0\cdot4880(x + 0\cdot1816)$$
$$= 4\cdot875 - 0\cdot488x.$$

In Table 37 the probits calculated by substituting the seven values of x in this equation are shown. These and the new estimate of H agree so closely with the expected values used in

TABLE 37.   Comparison of Observed and Expected
Lime Sulphur Uptakes

x	Y	P	$U = HP$	u
3	3·411	0·056	4·8	4·6
2	3·899	0·135	11·6	7·6
1	4·387	0·270	23·1	25·2
0	4·875	0·450	38·5	35·0
−1	5·363	0·642	55·0	63·1
−2	5·851	0·803	68·7	69·0
−3	6·339	0·910	77·9	68·5
−∞	—	1·000	85·6	90·2

$$S(u - U)^2 = 207\cdot92.$$

Table 36 that a further cycle of computations is unnecessary. Had the new estimates of H or Y been markedly different, a second cycle would have been calculated with them as expected values.

A test of significance of the discrepancies between the observed uptakes and the estimates from the probit equation can now be made. From each value of Y in Table 37, P is obtained (by Table I) and multiplied by the revised estimate of H to give an expected uptake U. The sigmoid curve shown in Fig. 21 has been plotted from these values of U. A sum of squares of differences between U and the observed uptake u is then formed, and, since each u is a mean for eight cells, multiplied by 8 to bring it to the same units as those of Table 35. The mean square, based on 5 degrees of freedom since three parameters have been estimated from the eight values of u, is $(8 \times 207\cdot9)/5 = 332\cdot6$, which is less than the error mean square in Table 35 and therefore gives no evidence of significant deviations from the fitted sigmoid curve. Had this conclusion not been obvious, the ratio of the two mean squares would have been tested for significance (Fisher and Yates, 1943, Table V), and had the discrepancies been significant, either the hypothesis of the normal sigmoid curve would have had to be discarded or the variances of all estimates would have had to be multiplied by a heterogeneity factor.

Since no heterogeneity has been found, standard errors of estimates can be calculated directly from equation (10·7). For example, the standard error of H is $\pm \sqrt{(380\cdot8 \times 0\cdot1029)} = \pm 6\cdot26$, so that the adjustment to H made as a result of the computations is only one-tenth of the standard error. Again, the last element of the matrix in equation (10·7) gives the variance of b, $380\cdot8 \times 0\cdot00002492$, so that $b = 0\cdot488 \pm 0\cdot097$.

Using the term ED 50 for the concentration of lime sulphur producing a 50 % reduction in uptake, the log ED 50 is seen to be

$$m = -0\cdot256 \pm 0\cdot338,$$

and the variance of m is calculated as

$$V(m) = \frac{1}{b^2}\{V(a') + 2(m - \bar{x})\,\mathrm{Cov}\,(a', b) + (m - \bar{x})^2\,V(b)\}$$

$$= \frac{380\cdot8}{b^2}\{0\cdot00006870 + 2 \times 0\cdot074 \times 0\cdot00001727$$

$$+ (0\cdot074)^2 \times 0\cdot00002492\}$$

$$= 380\cdot8 \times 0\cdot00007140/0\cdot2381$$

$$= 0\cdot1142.$$

In calculating fiducial limits for m, the considerations of § 19 arise once more. For the 5 % level of probability $t = 2\cdot02$, and

$$g = t^2 V(b)/b^2 = 0\cdot16;$$

this value of g is too large for approximate fiducial limits placed at $0\cdot338t$ on either side of m to be sufficiently exact, and the formula (4·7) must therefore be used. From equation (10·7),

$$V(5 - a') = 0\cdot026161,$$
$$V(b) = 0\cdot009489,$$
$$\mathrm{Cov}\,(5 - a', b) = 0\cdot006577,$$

whence the fiducial limits are found to be $0\cdot357$ and $-1\cdot161$.

Now equation (10·5) may be written

$$\log_{10} (\% \text{ concentration}) = 0.699x - 2.097,$$

from which the ED 50 of lime sulphur is estimated to be 0·0053 %, with 5 % fiducial limits at 0·0142 and 0·0012 %.

Many of the methods of analysis discussed in earlier chapters can be modified for use with quantitative response data of the type under consideration here. In particular, the estimation of the relative potency of two stimuli whose probit regression lines are parallel, the analysis of data involving two or more dose factors, and the study of mixtures, can all be effected by techniques analogous to those used with quantal responses. The chief differences are the change in the weighting coefficient and the frequent necessity of estimating the third parameter, H. Consequently the computations are more laborious than those of ordinary probit analysis, but they are by no means prohibitively so. Additional examples will not be given here, as the assumption of a linear dosage-response relationship is so often good enough; the probit procedure for many types of data should be apparent by analogy with the corresponding quantal response problems. McCallan (1943) has suggested a further modification, for use with certain types of quantitative data, in which an empirical relationship between weighting coefficient and response is estimated as a preliminary to the main analysis.

48. SEMI-QUANTAL RESPONSES

Intermediate between quantal responses and the truly quantitative responses are those which may be described as *semi-quantal*. As noted earlier, Tattersfield *et al.* (1925) classified their insects as dead, moribund, slightly affected, and unaffected, and thus recognized four levels of response instead of the two characteristic of quantal data. They reduced their results to quantal form, however, by assessing toxicity in terms of the percentage of insects which were either moribund or dead, and made no allowance for the subclassification of these insects into two levels or of the remainder into the two levels of unaffected and slightly affected; standard methods of probit analysis are then applicable to the data without modification.

Better use might be made of the data if the numbers of insects at the four different levels could be combined into a single index of toxic effect. Such a scheme was indeed proposed earlier by Fryer *et al.* (1923); they used the same fourfold classification and, in order to obtain a percentage response at each dose, scored moribund and slightly affected insects as one-half and one-quarter dead respectively. For example, if a batch of ten insects were classified as 4, 1, 3, 2, the proportionate response would be $(4 \times 1 + 1 \times 0.5 + 3 \times 0.25)/10$, or 52.5%.

This method of scoring the results did not appear to give very much smoother response curves than that subsequently adopted by Tattersfield, and has never been very widely used. Possibly discrimination between moribund and slightly affected is simpler than between other pairs of classifications, so that the information provided by the latter is relatively unreliable. Exact statistical methods appropriate to the analysis of semi-quantal data have not yet been developed, but they would certainly be much more complex than the ordinary probit analysis. On the other hand, probably a sufficiently good approximation for many practical purposes is to form an index such as that described in the previous paragraph, and to use this as the value of p in an ordinary probit analysis, but always to use an empirical variance for expressing the precision of estimates. The weights derived from equation (3.4) or its generalization (6.3) should give a reasonable indication of the *relative* value to be placed on different observations, but can no longer be regarded as the reciprocals of the true variances; the heterogeneity χ^2 obtained therefore cannot be considered a true χ^2, but is only a sum of squares from which an estimate of variance can be formed.

Appendix I

THE COMPUTING OF PROBIT ANALYSES

THE chief hindrances to the more widespread adoption of the probit method for the statistical analysis of quantal response data are probably the apparent complexity of the mathematical theory and the apparent laboriousness of the computations. The more technical details of the theory (Appendix II) are admittedly difficult, especially for the many biologists who lack mathematical training. In Chapters 2 and 3 an attempt has been made to present a reasonably simple account of the aim and underlying principles of the method, in a form which the reader will be able to appreciate even though he has to take the theory on faith. For many routine purposes the computations can be largely replaced by the graphical analysis described in Chapter 3, relative potencies and other parameters being estimated by measurement on the diagrams. When the complexity of the data, or the desirability of using extensions of the probit method (such as have been described in later chapters), makes necessary arithmetical rather than graphical estimation, the labour can be much reduced and computational accuracy much increased by orderly arrangement and systematic working procedure. In this appendix is given an example of the computing required for the simplest type of analysis, the fitting of a single probit regression line; the arrangement is that used throughout this book, but the steps are set out in greater detail and recommendations for adequate checking are made. With modifications appropriate to the various circumstances, the same procedure may be used in computing any of the more complicated probit regression equations that have been described.

The computer is assumed to be working with a calculating machine; the same results can be obtained by pen-and-paper calculations, but almost any machine intended for general computing will materially improve both accuracy and speed. Before buying a machine it is well to take expert advice on the most suitable type, though to some extent personal preference must decide between the products of different manufacturers. If the machine is to be used frequently, an electrically operated model is desirable, and a 10-figure rather than an 8-figure keyboard is an advantage that usually outweighs the additional cost.

The reader should find little difficulty in becoming proficient at the arithmetical processes described in this book, but the only route to accuracy and efficiency is by way of experience. A machine is not a complete safeguard against arithmetical errors, and carelessness will lead to wrong answers just as certainly as in non-mechanized calculations. One of the most frequent sources of error lies in copying from paper to machine or from machine to paper; inversions of the order of two digits and similar mistakes are particularly easily made. Computations should therefore be planned so as to reduce to a minimum the necessity for copying on to paper figures that have later to be restored to the machine. The surest means of preventing errors is to have all work checked by another computer on another machine, but, by the application of full checks at every stage, one computer should be able to carry out the work satisfactorily. The chief danger is that any misreading of figures, wrong setting of the machine, or faulty working of the machine may be repeated in checking; in order to guard against this, where a section of the computations cannot be checked by an independent path and the steps have therefore to be duplicated, the order of setting the machine should be changed so as to avoid the complete repetition of every detail. For example, a column may be summed by starting from the bottom instead of from the top, or a product found by interchanging the roles of multiplicand and multiplier.

Inexperienced computers frequently carry a far greater number of digits in their computations than is warranted by the accuracy of the original data, and present their results to six places of decimals when three at most are justifiable. In machine calculations an increase in the number of digits does not increase the labour to the same extent as in pen-and-paper work, and undoubtedly it is sometimes easier to carry an additional decimal place than to decide how many can be justified. Nevertheless the saving of time through working with fewer digits may be considerable in a long series of analyses; it should further be remembered that copying errors are less common with four digit numbers than with seven. The general practice in the numerical examples of this book has been to cut out all unnecessary digits from the early stages of an analysis (i.e. in dosages, working probits, weights, etc.), to retain digits fairly fully at intermediate stages (sums of squares and products, elements of matrices, etc.) as an aid to checking, but to present results shorn of superfluous digits and free of any spurious appearance of accuracy. Unless doses have been measured to within $0 \cdot 1 \%$ of their true values, two or three significant digits in the dosage, x, are sufficient. Percentage kills based on batches of 200 individuals or less need not be expressed more precisely than to the nearest 1 %, thus enabling the tables of this book to be used without interpolation; there is little to be gained by calculating the kills to greater accuracy than the nearest 0·1 % unless the batches contain more than 2000 individuals.

Martin (1940, Table 3) has given data on the toxicity to *Aphis rumicis* of an ether extract of *Derris malaccensis*; the original observations of the experiment have been reconstructed from his figures in order that they may be used in an example of the typical probit analysis. Table 38 shows the greater portion of the computations for estimating a linear regression equation which will relate the probit mortality to the log concentration.

TABLE 38.   Computations for the Fitting of a Probit
Regression Equation

λ	x	n	r	p'	$p(C=2)$	Empirical probit	Y	nw	y	nwx	nwy
520	2·72	49	49	100	100	∞	7·5	2·4	7·85	6·528	18·840
390	2·59	45	44	98	98	7·05	6·9	6·8	7·03	17·612	47·804
260	2·41	44	38	86	86	6·08	6·1	17·4	6·08	41·934	105·792
130	2·11	52	15	29	28	4·42	4·7	30·4	4·43	64·144	134·672
65	1·81	47	4	9	7	3·52	3·3	6·7	3·57	12·127	23·919
0	—	48	1	2	0	—	—	—	—	—	—
								63·7		142·345	331·027

$1/Snw = 0.0156986$, $\bar{x} = 2.2346$, $\bar{y} = 5.1967$.

$Snwx^2$	$Snwxy$	$Snwy^2$
321·72589	757·4672	1809·159
318·08633	739·7180	1720·234
3·63956	17·7492	88·925
		86·558
		$2.37 = \chi^2_{[3]}$

The various steps in building up Table 38 and in completing the estimation of the probit regression line are as follows:

1. In the column headed λ enter, in suitable units (here milligrams of dry root per litre of spray fluid), the doses tested, arranging these in descending order from the highest to the controls or zero concentration.

2. In the column headed x enter the logarithms of λ, to base 10, correct to two places of decimals. The doses may be multiplied or divided by a power of 10 throughout in order to make x take small

positive values, but a compensating adjustment must be made in later stages of the analysis if the results are to be expressed in the original units of dose.

3. In the columns headed n and r enter for each dose the number of insects tested and the number badly affected, moribund, or dead (Martin's $B + M + D$, which he takes as the 'kill').

4. Check that steps 1, 2 and 3 are correct, beginning each check from the bottom of the column.

5. Calculate the percentage kill, $p' = 100r/n$, to the nearest whole number; this step is most expeditiously performed on a slide rule. If n exceeds 200 for many of the doses, give the percentages to one decimal place.

6. The percentage kill among the controls is small $(c = 2)$, and is estimated from about as many insects as the kill for other doses; the approximate analysis discussed in § 27 will therefore be sufficiently accurate. The population value, C, is taken as equal to c, and the adjusted kills calculated accordingly as

$$p = \frac{p' - 2}{98} \times 100$$

$$= 1 \cdot 020p' - 2 \cdot 0;$$

p also is taken correct to the nearest whole number (or to 1 decimal place if n is generally greater than 200). A slide rule can be used for the multiplication.

7. Check step 5, multiplying p' by n to give $100r$.

8. Check step 6 from the relation

$$p' = 0 \cdot 98p + 2.$$

9. Enter the probits of p in the 'empirical probit' column, reading the values from Table I or Table 1, to 2 decimal places.

10. Plot the empirical probits against x, as shown in Fig. 23; draw a provisional straight line to fit the points, placing the line by eye and allowing for one point above the line at $x = 2 \cdot 72$.

11. For each of the dosages used in the experiment read the value of the ordinate to the provisional line. These are the expected probits, Y, and are entered in the appropriate column of Table 38, correct to 1 decimal place; greater accuracy in Y is unnecessary unless n is very large.

12. Check steps 9, 10 and 11.

13. From the column $C = 2$ of Table II, read the weighting coefficient for each Y in Table 38, multiply by the corresponding n, and enter, to 1 place of decimals, in the column nw. Two or three significant digits in nw will generally be sufficient, and a good working rule is to use 2 places of decimals when n is less than 20, 1 place when n is between 20 and 200, and the nearest whole number for higher values

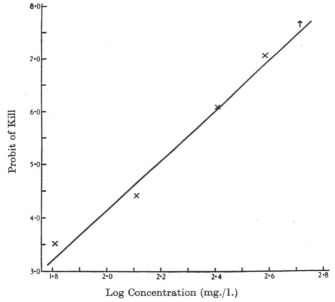

FIG. 23. Relationship between dosage of *D. malaccensis* and probit of kill of *A. rumicis*, showing probit regression line.

of n, modifying this rule so as to give the same number of decimal places for each dose throughout one analysis. In exceptional circumstances, such as the occurrence of a very wide discrepancy between the provisional line and an observation at either extreme of the range, a greater number of decimal places in nw may be desirable for the one dose. These recommendations assume that C is not very large; more decimal places may be required to give a reasonable number of significant digits when a high value of C reduces the values of w considerably.

14. From Table IV enter the working probit, y, corresponding to each p (not p') and Y. Thus from the last page of this table, the last

line of the column headed 7·5 gives $y = 7·85$ for $p = 100$; from the preceding page the column headed 6·9 gives $y = 7·03$ for $p = 98$, and so on. Use 2 decimal places if n is generally less than 200, otherwise 3. If Y is less than 2·0 or greater than 7·9, determine y from Table III as described in § 16.

15. Check steps 13 and 14; 13 may be checked satisfactorily with a slide rule. It will be observed that y usually, but not always, differs less from the empirical probit than does Y.

16. Place the first value of nw on the keyboard of the machine, multiply by the corresponding x, and enter the product in the column nwx; leave the keyboard unaltered but clear the product from the machine, multiply by y, and enter in the column nwy. Clear the machine and repeat for each dosage, entering each result to its full number of decimal places. No entries are made in the line for $\lambda = 0$, as the controls are used only for estimating C.

17. Place the last value of x on the right-hand side of the keyboard, unity at the left-hand side, and multiply by nw; without clearing the result, repeat with the next to the last value of x and so work to the top of the x column, thus accumulating the products at the right-hand side of the result register and the sum of nw at the left hand. Enter the totals

$$Snw \; = \; 63·7,$$
$$Snwx = 142·345,$$

in Table 38, and check the latter by addition of the nwx column from the top. Before $Snwx$ is cleared from the machine, divide it by Snw to give $\bar{x}$ to 4 places of decimals, and enter $\bar{x}$ in Table 38.

18. Repeat these operations using y in place of x, thus checking the total Snw and also obtaining

$$Snwy = 331·027;$$

again check, by adding the nwy column from the top, and before clearing divide by 63·7 to give $\bar{y}$ to 4 places of decimals. Enter $\bar{y}$ in Table 38.

19. Find the reciprocal of Snw, either from tables or by division, and enter at the bottom of Table 38 to at least 7 decimal places. Additional accuracy is needed here, as this quantity is to be multiplied by large numbers.

20. Set $1/Snw$ on the keyboard and check that multiplication by Snw gives unity; clear the result and multiply by $Snwx$, so checking the values of $\bar{x}$; clear the result and similarly check $\bar{y}$.

21. Square $Snwx$ and divide by Snw, multiply $Snwx$ and $Snwy$ and divide by Snw, square $Snwy$ and divide by Snw; the answers 318·08633, 739·7180 and 1720·234, are to be entered in Table 38 as shown. Check by squaring the second, dividing by the first, and obtaining the third as the quotient; the first being still on the keyboard, multiply by 63·7 and divide by 142·345 to give 142·345 as a quotient. These checks may fail in the last digit as a result of rounding off. The parts played in the checks by the first and third of the three calculated quantities should be interchanged when the third has the greater number of significant digits.

22. Set the first entry for nwx on the keyboard, multiply by x, and repeat with successive lines of Table 38, accumulating the total so as to give

$$Snwx^2 = 6\cdot528 \times 2\cdot72 + 17\cdot612 \times 2\cdot59 + \ldots + 12\cdot127 \times 1\cdot81$$

$$= 321\cdot72589.$$

Enter this figure in the appropriate position in the lower part of the table. Multiply the last value in the nwx column, which is already on the keyboard, by y and repeat with successive entries up the table, to obtain

$$Snwxy = 757\cdot4672;$$

enter this figure. Multiply nwy by x $down$ the table, in order to check $Snwxy$. Before clearing the machine, subtract from $Snwxy$ the second of the three quantities calculated in step 21, without the machine, to give

$$S_{xy} = 17\cdot7492;$$

enter in Table 38 and check the difference by means of the machine. Multiply nwy by y up the table to give $Snwy^2$, and enter in Table 38.

23. Check $Snwx^2$ by accumulating products of nwx and x up the table. Before clearing, subtract the first of the quantities calculated in step 21, without the machine, to give

$$S_{xx} = 3\cdot63956;$$

enter in the table and check with the machine. Similarly check $Snwy^2$ $down$ the table, derive

$$S_{yy} = 88\cdot925,$$

enter in the table, and check.

24. Compute

$$b = 17\cdot7492 \div 3\cdot63956$$

$$= 4\cdot8767.$$

25. Obtain the equation for the probit regression line as

$$Y = 5 \cdot 1967 + 4 \cdot 8767(x - 2 \cdot 2346)$$
$$= -5 \cdot 70 + 4 \cdot 88x;$$

3 places of decimals are almost always sufficient for the coefficients of this equation, and consideration of the standard errors to which the parameters are subject often suggests, as here, that only 2 are justified.

26. Compute $(17 \cdot 7492)^2 \div 3 \cdot 63956$, and subtract the result from S_{yy}, giving
$$\chi^2_{[3]} = 2 \cdot 37;$$
reference to Table VI shows this not to be indicative of significant heterogeneity, and variances may therefore be derived from true weights without any heterogeneity factor. When a significant χ^2 is obtained at this stage, modify subsequent steps as in Exs. 7 and 9.

27. To check b, set $3 \cdot 63956$ and multiply by $4 \cdot 8767$ to obtain $17 \cdot 7492$.

28. The variance of b is $1/S_{xx}$. Find the standard error of b as the square root of this reciprocal, and check by calculating the reciprocal of $\sqrt{S_{xx}}$; a slide rule or tables of square roots and reciprocals may be used. The result is
$$b = 4 \cdot 88 \pm 0 \cdot 52,$$
and the size of the standard error shows that there is no need to quote b to more than 2 decimal places.

29. Check step 25. Calculate values of Y for three values of x (say $x = 1 \cdot 5$, $2 \cdot 0$ and $2 \cdot 5$), and plot Y against x on Fig. 23. The three points should be collinear and they define the probit regression line. This line is almost indistinguishable from the provisional line and has therefore not been drawn separately in the figure. Since the agreement is so good, there is no need to carry out a second cycle of computations.

30. Find the log LD50 as the value of x which gives $Y = 5$:
$$m = (5 + 5 \cdot 70)/4 \cdot 88$$
$$= 2 \cdot 193.$$
The antilogarithm of m is the estimated LD50, 156 mg./l.

31. Calculate
$$V(m) = \frac{1}{(4 \cdot 877)^2} \left\{ 0 \cdot 0157 + \frac{(2 \cdot 193 - 2 \cdot 235)^2}{3 \cdot 640} \right\}$$
$$= (0 \cdot 0157 + 0 \cdot 0005)/23 \cdot 79$$
$$= 0 \cdot 00068,$$
whence the standard error of m is $\pm 0 \cdot 026$.

32. Calculate $g = t^2 V(b)/b^2$ for the probability level at which fiducial limits are required; for the 5 % level $t = 1.96$ (Table VII), so that
$$g = (1.96)^2/3.640 \times 23.79$$
$$= 0.044.$$

33. Since g is small, the fiducial limits to m may be taken at a distance 1.96×0.026 on either side of m, these therefore being 2·244 and 2·142. The corresponding concentrations are 175 and 139 mg./l. respectively.*

34. Check steps 30–33.

35. The conclusion from the analysis of these data is that the median lethal dose of Martin's ether extract of *Derris malaccensis* to *Aphis rumicis*, under the conditions of the experiment, has a maximum likelihood estimate of 156 mg./l.; with a fiducial probability of 95 %, the true value of the median lethal dose may be expected to lie between 175 and 139 mg./l.

The above is a detailed account of a systematic arrangement of the computations suitable for use when a single computer is responsible for the whole analysis. The plan need not be followed exactly and no doubt personal taste will suggest modifications. There is necessarily a conflict between the desirability of checking a result before too many further calculations have been based upon it and the desirability of delaying a check for some time so as to reduce the risk of unconscious repetition of mistakes. The computer's aim should be to carry out the original calculations correctly, and checking should be looked upon as a verification of correctness rather than as means of discovering errors. Where a result can be checked independently of its first calculation the check may be made immediately, but where the only check is to repeat the same processes a reasonable interval should be allowed to elapse first. In the arrangement of the computations that has just been described, notes on checks have been inserted, but generally the reader has been left to devise his own methods for these; they should not be made earlier than the points indicated, and preferably the original working should be so accurate that they can safely be left until much later. When a second computer is available to assist in the analysis, he should be responsible for all checks.

* Fiducial limits to m calculated from the exact equation (4·6) are 2·243 and 2·139, and the concentrations are 175 and 138 mg./l. respectively, thus confirming the statement that g is too small for the exact formula to be needed.

The reader who adopts either this or a similar arrangement of the work for the simple type of data should have no difficulty in adapting and extending it to the more complex analyses discussed in Chapters 5–10. For example, when relative potencies are being estimated for several poisons tested in a single experiment, the provisional lines (step 10) will be drawn parallel if there seems any possibility of data agreeing satisfactorily with the hypothesis that the probit regression lines have equal slopes. Thereafter the computations proceed as outlined above, for each poison separately, until the values of S_{xx}, S_{xy} and S_{yy} have been obtained for each. A test of parallelism and, if departures from parallelism are not significant, subsequent estimation of relative potencies based on a common regression coefficient follow as described in § 20. The computations are analogous to those in steps 24–34 above.

Appendix II

MATHEMATICAL BASIS OF THE PROBIT METHOD

In Chapters 1–10 of this book an attempt has been made to develop the practical and computational aspects of probit analysis while giving little more than a hint of the theory underlying the processes of estimation. Most of those who are concerned with the use of the probit transformation in the analysis of numerical data will be content to take on faith the mathematical framework. This appendix has been written for the benefit of the mathematically minded reader, as a brief outline of the derivation of the equations for estimating the parameters which have been used in other parts of the text. The whole theory of the probit method is derived from the principles outlined in this appendix, and the mathematical statistician should have little difficulty in discovering for himself how the equations given here lead to the computational procedure used in earlier chapters.

If a batch of n insects is exposed to a poison at a dose sufficient to produce an average kill of a proportion P, and if the insects react independently of one another, then the probability of r deaths is (§ 7):

$$\frac{n!}{r!(n-r)!} P^r Q^{n-r}.$$

Suppose now that a series of k doses is tested in an experiment; then the probability of a particular set of numbers killed in each group is proportional to e^L, where

$$L = Sr \log P + S(n-r) \log Q, \qquad (\text{II}, 1)$$

and S denotes summation over all doses. The quantity e^L, or, more strictly, a quantity proportional to it but having a maximum value of unity, has been called by Fisher (1922) the *Likelihood* of the observations.

Now P, Q ($= 1 - P$) are functions of the dose which contain certain parameters, and the problem confronting the statistician is that of estimating the parameters from the experimental data. Fisher (1922, 1925) has shown that estimates of the parameters which maximize the likelihood are *efficient* in the sense of having minimal sampling variance in large samples. The likelihood is a maximum when L is

14

a maximum; hence, if θ is a parameter of the distribution of individual tolerances, the maximum likelihood estimate of θ must satisfy the equation

$$0 = \frac{\partial L}{\partial \theta} = S\frac{r}{P}\frac{\partial P}{\partial \theta} + S\frac{n-r}{Q}\frac{\partial Q}{\partial \theta}$$

$$= S\frac{n(p-P)}{PQ}\frac{\partial P}{\partial \theta}, \tag{II, 2}$$

where $p = r/n$ is an estimate of P. If more than one parameter requires to be estimated, a set of equations such as (II, 2) must be satisfied simultaneously.

Direct solution of these equations is frequently impossible, but they may easily be solved by a process of successive approximations. For suppose that θ, ϕ are parameters to be estimated and that θ_1, ϕ_1 have been obtained, in any way, as first approximations to the maximum likelihood estimates. By the Taylor-Maclaurin expansion, to the first order of small quantities

$$\left.\begin{aligned}
\frac{\partial L}{\partial \theta_1} + \delta\theta\frac{\partial^2 L}{\partial \theta_1^2} + \delta\phi\frac{\partial^2 L}{\partial \theta_1 \partial \phi_1} = 0,\\[2mm]
\frac{\partial L}{\partial \phi_1} + \delta\theta\frac{\partial^2 L}{\partial \theta_1 \partial \phi_1} + \delta\phi\frac{\partial^2 L}{\partial \phi_1^2} = 0,
\end{aligned}\right\} \tag{II, 3}$$

where the addition of the suffix to θ, ϕ implies that the first approximations are substituted after differentiation. The quantities $\frac{\partial^2 L}{\partial \theta_1^2}$, $\frac{\partial^2 L}{\partial \theta_1 \partial \phi_1}$, and $\frac{\partial^2 L}{\partial \phi_1^2}$ may be replaced by their expectations,[*] obtained by putting $p = P$ after differentiation; $\delta\theta$, $\delta\phi$, adjustments to θ, ϕ are then given by the equations

$$\left.\begin{aligned}
\delta\theta S\frac{n}{P_1 Q_1}\left(\frac{\partial P_1}{\partial \theta_1}\right)^2 + \delta\phi S\frac{n}{P_1 Q_1}\frac{\partial P_1}{\partial \theta_1}\frac{\partial P_1}{\partial \phi_1} = S\frac{n(p-P_1)}{P_1 Q_1}\frac{\partial P_1}{\partial \theta_1},\\[2mm]
\delta\theta S\frac{n}{P_1 Q_1}\frac{\partial P_1}{\partial \theta_1}\frac{\partial P_1}{\partial \phi_1} + \delta\phi S\frac{n}{P_1 Q_1}\left(\frac{\partial P_1}{\partial \theta_1}\right)^2 = S\frac{n(p-P_1)}{P_1 Q_1}\frac{\partial P_1}{\partial \phi_1}.
\end{aligned}\right\} \tag{II, 4}$$

[*] Garwood (1940) has developed the maximum likelihood solution to the fundamental probit problem in a way very similar to that used here. He has also compared numerical results obtained by using expected values of the second differential coefficients (as here) with those obtained by using values calculated from the observations. Both processes must converge to the same result, but Garwood finds the second to converge a little more rapidly; in practice, however, the gain is more than counter-balanced by the greater time required for computing each cycle. The tables included in this book make the computations for the first method very much less onerous than those needed when the 'observed' second differential coefficients have to be calculated.

A second pair of adjustments is obtained by recalculating the equations with the new approximations, $\theta_1 + \delta\theta$ and $\phi_1 + \delta\phi$, in place of θ_1 and ϕ_1. The process may be repeated until the last set of adjustments is negligible; if some care is exercised in the choice of the first approximations, one or two cycles will usually suffice to give a reasonable numerical accuracy for practical purposes.

Fisher (1922, 1944, § 29) has further pointed out the important conclusion for all maximum likelihood estimation that, if $\breve{\theta}$ is the estimate of a single parameter obtained from equation (II, 2), the variance is

$$V(\breve{\theta}) = -1 \bigg/ \frac{\partial^2 L}{\partial \breve{\theta}^2};$$

$\breve{\theta}$ is to be substituted for θ after the differentiation. When more than one parameter is to be estimated, the variances and covariances are obtained from the inverse of the matrix of coefficients of $\delta\theta$, $\delta\phi$ in equations (II, 3). The result for two parameters is

$$V = \begin{pmatrix} -\dfrac{\partial^2 L}{\partial \breve{\theta}^2} & -\dfrac{\partial^2 L}{\partial \breve{\theta} \partial \phi} \\ -\dfrac{\partial^2 L}{\partial \breve{\theta} \partial \phi} & -\dfrac{\partial^2 L}{\partial \phi^2} \end{pmatrix}^{-1}.$$

In practice, the matrix with the last approximations (θ_1, ϕ_1 instead of $\breve{\theta}$, $\breve{\phi}$) is inverted as part of the last cycle of computations, and the variances and covariances are taken as approximately equal to the elements of this inverse. A χ^2 test for the heterogeneity of the departures of the observations from the dose-response law specified by the estimated parameters then follows. Examples of this technique for studying the errors of estimation and the goodness of fit have occurred in §§ 28, 31 and 47.

The equations are of general applicability, whatever the form of P, and may easily be extended to the estimation of a greater number of parameters. Toxicity test data and other quantal response problems require their simplification to more specialized forms suitable for computation. The simplest of these concerns the estimation of the parameters of the tolerance distribution given by equation (2·6), or

$$P = \frac{1}{\sigma\sqrt{(2\pi)}} \int_{-\infty}^{x} e^{-\frac{(x-\mu)^2}{2\sigma^2}} \, dx,$$

where x measures the dosage on a logarithmic or other normalizing scale. This distribution has been shown in § 9 to be equivalent to a linear relationship between x and the probit of P, say

$$Y = \alpha + \beta x,\qquad\qquad (II,5)$$

where Y is defined by equation (3·1), namely

$$P = \frac{1}{\sqrt{(2\pi)}} \int_{-\infty}^{Y-5} e^{-\frac{1}{2}u^2}du.$$

Now

$$\frac{\partial P}{\partial Y} = \frac{1}{\sqrt{(2\pi)}} e^{-\frac{1}{2}(Y-5)^2} = Z,$$

the ordinate to the normal curve at the point whose abscissa is ($Y-5$), and therefore

$$\frac{\partial P}{\partial \alpha} = Z, \quad \frac{\partial P}{\partial \beta} = Zx.$$

Hence, if

$$Y = a + bx$$

is a first approximation to the maximum likelihood estimate of equation (II, 5), adjustments to a and b are given by

$$\delta a S\frac{nZ^2}{PQ} + \delta b S\frac{nZ^2}{PQ}x = S\frac{nZ^2}{PQ}\left(\frac{p-P}{Z}\right),$$

$$\delta a S\frac{nZ^2}{PQ}x + \delta b S\frac{nZ^2}{PQ}x^2 = S\frac{nZ^2}{PQ}\left(\frac{p-P}{Z}\right)x.$$

These equations are most readily solved by introducing the weighting coefficient, $w = Z^2/PQ$, writing $\bar{x}$ for the weighted mean dosage $Snwx/Snw$, and putting the provisional equation in the form

$$Y = a' + b(x - \bar{x}),\qquad\qquad (II,6)$$

where

$$a' = a + b\bar{x}.$$

The equations for $\delta a'$, δb then reduce to

$$\delta a' Snw = Snw\left(\frac{p-P}{Z}\right),$$

$$\delta b Snw(x - \bar{x})^2 = Snw(x-\bar{x})\left(\frac{p-P}{Z}\right).$$

It follows that, if the *working probit*, y, is defined by

$$y = Y + \frac{p-P}{Z},\qquad\qquad (II,7)$$

Y, P and Z being determined from the first approximations to the parameters, a new approximation to the regression equation (II, 5) has

$$a' = \bar{y} = Snwy/Snw,\qquad\qquad (II,8)$$

$$b = Snw(x-\bar{x})(y-\bar{y})/Snw(x-\bar{x})^2,\qquad\qquad (II,9)$$

and is thus obtained as the weighted linear regression of y on x. This method of solution of the maximum likelihood equations was first given by Fisher (1935). The variances of $\bar{y}$, b are, by the ordinary formulae for a weighted linear regression,

$$V(\bar{y}) = 1/Snw, \tag{II, 10}$$

$$V(b) = 1/Snw(x-\bar{x})^2, \tag{II, 11}$$

and the same values may be obtained by matrix inversion.

Homogeneity of the experimental material, as evidenced by departures from the fitted probit regression equation, has so far been assumed. If in fact the n test organisms in a batch do not react independently to the dose applied, but depart from the fitted equation (II, 6) more markedly than can be attributed to random variation, without showing any regularity such as would suggest that a more complex equation is required, the weights of the observations must be decreased, and the variances correspondingly increased. This situation will in general be indicated by a significantly large value of the heterogeneity χ^2, whose calculation is described in § 18. Provided that no classes have had to be amalgamated (cf. Ex. 7), χ^2 will have $(k-2)$ degrees of freedom, two less than the number of batches tested; $\chi^2/(k-2)$ is then the factor by which the variances must be increased. In more complex analyses the heterogeneity factor remains the mean square derived from the heterogeneity χ^2. When a heterogeneity factor has to be used, all variances and standard errors must be considered in relation to a t-distribution, not a normal, with degrees of freedom equal to those of χ^2. This procedure is not a method peculiar to probit analysis, but is the normal statistical practice of using an empirical variance, instead of a theoretical, when the data give evidence of the occurrence of variation other than that included in the theoretical value.

As an illustration of the application of equations (II, 3) to a more complex situation, it is instructive to consider in detail the problem of § 28. When the mortality due to the poison is supplemented by an average natural mortality, C, the expected death-rate is

$$P' = C + P(1-C).$$

The quantities P', Q' replace P, Q in (II, 1), and the prototype of the maximum likelihood equations is

$$\frac{\partial L}{\partial \theta} = S\frac{n(p'-P')}{P'Q'}\frac{\partial P'}{\partial \theta} = 0.$$

Now
$$\frac{\partial P'}{\partial C} = Q, \quad \frac{\partial P'}{\partial P} = (1 - C),$$

and also
$$p' - P' = (p - P)(1 - C).$$

Hence the equations become

$$\frac{\partial L}{\partial C} = \frac{n_c(c - C)}{C(1 - C)} + S\frac{n(p - P)}{C + P(1 - C)} = 0,$$

$$\frac{\partial L}{\partial \alpha} = S\frac{nZ^2(1 - C)}{Q\{C + P(1 - C)\}}\left(\frac{p - P}{Z}\right) = 0,$$

$$\frac{\partial L}{\partial \beta} = S\frac{nZ^2(1 - C)}{Q\{C + P(1 - C)\}}\left(\frac{p - P}{Z}\right)x = 0,$$

where S is used for summation over all doses, omitting the controls which show a mortality rate c in a total number n_c. Write

$$w = \frac{Z^2}{Q\left\{P + \dfrac{C}{1 - C}\right\}},$$

as in equation (6·3), differentiate once more and equate observed values with expected; the following are then obtained as coefficients in the equations (II, 3), which give adjustments δC, δa and δb to provisional values of the parameters C, a and b:

$$-\frac{\partial^2 L}{\partial C^2} = \frac{n_c}{C(1 - C)} + \frac{1}{(1 - C)^2}Snw\frac{Q^2}{Z^2},$$

$$-\frac{\partial^2 L}{\partial C \partial \alpha} = \frac{1}{1 - C}Snw\frac{Q}{Z},$$

$$-\frac{\partial^2 L}{\partial C \partial \beta} = \frac{1}{1 - C}Snwx\frac{Q}{Z},$$

$$-\frac{\partial^2 L}{\partial \alpha^2} = Snw,$$

$$-\frac{\partial^2 L}{\partial \alpha \partial \beta} = Snwx,$$

$$-\frac{\partial^2 L}{\partial \beta^2} = Snwx^2.$$

If the regression equation is rewritten as in (II, 6), and the working probit is again introduced, the equations for δC, a' and b may be reduced to the form most suitable for computing, namely that of equations (6·4). Since these three equations are not independent of one another, the variances of the estimated parameters can only be obtained by inversion of the matrix of coefficients in the manner illustrated in Ex. 17.

The estimates of the parameters for any other form of the probability function, P, can be derived in a similar way, and equations suitable for solution by successive approximations will be obtained. The reader will find it instructive to develop the equations for the Parker-Rhodes hypothesis, equations (9·11), by this method.

REFERENCES

ABBOTT, W. S. (1925). A method of computing the effectiveness of an insecticide. *J. Econ. Ent.* **18**, 265–7.

BARTLETT, M. S. (1937). Some examples of statistical methods of research in agriculture and applied biology. *J. R. Statist. Soc. Suppl.* **4**, 137–83.

BEHRENS, B. (1929). Zur Auswertung der Digitalisblätter im Froschversuch. *Arch. exp. Path. Pharmak.* **140**, 237–56.

BERKSON, J. (1944). Application of the logistic function to bio-assay. *J. Amer. Statist. Ass.* **39**, 357–65.

BLISS, C. I. (1934a). The method of probits. *Science*, **79**, 38–9.

BLISS, C. I. (1934b). The method of probits—a correction. *Science*, **79**, 409–10.

BLISS, C. I. (1935a). The calculation of the dosage-mortality curve. *Ann. Appl. Biol.* **22**, 134–67.

BLISS, C. I. (1935b). The comparison of dosage-mortality data. *Ann. Appl. Biol.* **22**, 307–33.

BLISS, C. I. (1935c). Estimating the dosage-mortality curve. *J. Econ. Ent.* **28**, 646–7.

BLISS, C. I. (1938). The determination of dosage-mortality curves from small numbers. *Quart. J. Pharm.* **11**, 192–216.

BLISS, C. I. (1939a). The toxicity of poisons applied jointly. *Ann. Appl. Biol.* **26**, 585–615.

BLISS, C. I. (1939b). Fly spray testing: A discussion on the theory of evaluating liquid household insecticides by the Peet-Grady method. *Soap*, **15**, no. 4, 103–11.

BLISS, C. I. (1940a). Factorial design and covariance in the biological assay of vitamin D. *J. Amer. Statist. Ass.* **35**, 498–506.

BLISS, C. I. (1940b). The relation between exposure time, concentration and toxicity in experiments on insecticides. *Ann. Ent. Soc. Amer.* **33**, 721–66.

BLISS, C. I. (1941). Biometry in the service of biological assay. *Industr. Engng Chem.* (Anal. ed.), **13**, 84–8.

BLISS, C. I. and CATTELL, McK. (1943). Biological assay. *Ann. Rev. Physiol.* **5**, 479–539.

BLISS, C. I. and HANSON, J. C. (1939). Quantitative estimation of the potency of digitalis by the cat method in relation to secular variation. *J. Amer. Pharm. Ass.* **28**, 521–30.

BLISS, C. I. and MARKS, H. P. (1939a). The biological assay of insulin. I. Some general considerations directed to increasing the precision of the curve relating dosage and graded response. *Quart. J. Pharm.* **12**, 82–110.

BLISS, C. I. and MARKS, H. P. (1939b). The biological assay of insulin. II. The estimation of drug potency from a graded response. *Quart. J. Pharm.* **12**, 182–205.

BLISS, C. I. and PACKARD, C. (1941). Stability of the standard dosage-effect curve for radiation. *Amer. J. Roentgenol.* **46**, 400–4.

BLISS, C. I. and ROSE, C. L. (1940). The assay of parathyroid extract from the serum calcium of dogs. *Amer. J. Hyg.* **31**, A 79–98.

BROWN, W. and THOMSON, G. H. (1940). *The Essentials of Mental Measurement* (4th ed.). London: Cambridge University Press.

BURN, J. H. (1937). *Biological Standardization.* London: Oxford University Press.

BUSVINE, J. R. (1938). The toxicity of ethylene oxide to *Calandra oryzae*, *C. granaria*, *Tribolium castaneum*, and *Cimex lectularius*. *Ann. Appl. Biol.* **25**, 605–32.

BUTLER, C. G., FINNEY, D. J. and SCHIELE, P. (1943). Experiments on the poisoning of honeybees by insecticidal and fungicidal sprays used in orchards. *Ann. Appl. Biol.* **30**, 143–50.

CAMPBELL, F. L. (1930). A comparison of four methods for estimating the relative toxicity of stomach poison insecticides. *J. Econ. Ent.* **23**, 357–70.

CAMPBELL, F. L. and FILMER, R. S. (1929). A quantitative method of estimating the relative toxicity of stomach-poison insecticides. *Int. Congr. Ent.* **4**, 523–33.

CAMPBELL, F. L. and MOULTON, F. R. (editors) (1943). *Laboratory Procedures in Studies of the Chemical Control of Insects.* Washington, D.C.: Amer. Ass. Adv. Sci.

CLARK, A. J. (1933). *The Mode of Action of Drugs upon Cells.* London: Edward Arnold and Co.

CLOPPER, C. J. and PEARSON, E. S. (1934). The use of confidence or fiducial limits illustrated in the case of the binomial. *Biometrika*, **26**, 404–13.

COCHRAN, W. G. (1938). Appendix to a paper by TATTERSFIELD, F. and MARTIN, J. T. *Ann. Appl. Biol.* **25**, 426–9.

COCHRAN, W. G. (1942). The χ^2 correction for continuity. *Iowa St. Coll. J. Sci.* **16**, 421–36.

COWARD, K. H. (1938). *The Biological Standardization of the Vitamins.* London: Baillière, Tindall and Cox.

DE BEER, E. J. (1941). A scale for graphically determining the slopes of dose-response curves. *Science*, **94**, 521–2.

DE BEER, E. J. (1945). The calculation of biological assay results by graphic methods. The all-or-none type of response. *J. Pharmacol.* **85**, 1–13.

DIMOND, A. E., HORSFALL, J. G., HEUBERGER, J. W. and STODDARD, E. M. (1941). Role of the dosage-response curve in the evaluation of fungicides. *Bull. Conn. Agric. Exp. Sta.* no. 451.

DURHAM, F. M., GADDUM, J. H. and MARCHAL, J. E. (1929). Reports on biological standards. II. Toxicity tests for Novarsenobenzene (Neosalvarsan). *Spec. Rep. Ser. Med. Res. Coun., Lond.*, no. 128.

EPSTEIN, B. and CHURCHMAN, C. W. (1944). On the statistics of sensitivity data. *Ann. Math. Statist.* **15**, 90–6.

FECHNER, G. T. (1860). *Elemente der Psychophysik.* Leipzig: Breitkopf und Härtel.

FERGUSON, G. A. (1942). Item selection by the constant process. *Psychometrika*, 7, 19–29.

FIELLER, E. C. (1940). The biological standardization of insulin. *J.R. Statist. Soc. Suppl.* 7, 1–64.

FIELLER, E. C. (1944). A fundamental formula in the statistics of biological assay, and some applications. *Quart. J. Pharm.* 17, 117–23.

FINNEY, D. J. (1942a). The analysis of toxicity tests on mixtures of poisons. *Ann. Appl. Biol.* 29, 82–94.

FINNEY, D. J. (1942b). Examples of the planning and interpretation of toxicity tests involving more than one factor. *Ann. Appl. Biol.* 29, 330–2.

FINNEY, D. J. (1943a). The statistical treatment of toxicological data relating to more than one dosage factor. *Ann. Appl. Biol.* 30, 71–9.

FINNEY, D. J. (1943b). The design and interpretation of bee experiments. *Ann. Appl. Biol.* 30, 197.

FINNEY, D. J. (1944a). The application of the probit method to toxicity test data adjusted for mortality in the controls. *Ann. Appl. Biol.* 31, 68–74.

FINNEY, D. J. (1944b). Mathematics of biological assay. *Nature, Lond.*, 153, 284.

FINNEY, D. J. (1944c). The application of probit analysis to the results of mental tests. *Psychometrika*, 9, 31–9.

FINNEY, D. J. (1945). Microbiological assay of vitamins: The estimate and its precision. *Quart. J. Pharm.* 18, 77–82.

FISHER, R. A. (1922). On the mathematical foundations of theoretical statistics. *Philos. Trans.* A, 222, 309–68.

FISHER, R. A. (1925). Theory of statistical estimation. *Proc. Camb. Phil. Soc.* 22, 700–25.

FISHER, R. A. (1935). Appendix to BLISS, C. I.: The case of zero survivors. *Ann. Appl. Biol.* 22, 164–5.

FISHER, R. A. (1942). *The Design of Experiments* (3rd ed.). Edinburgh: Oliver and Boyd.

FISHER, R. A. (1944). *Statistical Methods for Research Workers* (9th ed.). Edinburgh: Oliver and Boyd.

FISHER, R. A. and YATES, F. (1943). *Statistical Tables for Biological, Agricultural and Medical Research* (2nd ed.). Edinburgh: Oliver and Boyd.

FRYER, J. C. F., STENTON, R., TATTERSFIELD, F. and ROACH, W. A. (1923). A quantitative study of the insecticidal properties of *Derris elliptica* (tuba root). *Ann. Appl. Biol.* 10, 18–34.

GADDUM, J. H. (1933). Reports on biological standards. III. Methods of biological assay depending on a quantal response. *Spec. Rep. Ser. Med. Res. Coun., Lond.*, no. 183.

GALTON, F. (1879). The geometric mean in vital and social statistics. *Proc. Roy. Soc.* 29, 365–7.

GARWOOD, F. (1940). The application of maximum likelihood to dosage-mortality curves. *Biometrika*, 32, 46–58.

GRIDGEMAN, N. T. (1943). The technique of the biological vitamin A assay. *Biochem. J.* 37, 127–32.

GRIDGEMAN, N. T. (1944). Mathematics of biological assay. *Nature, Lond.*, 153, 461–2.

HAZEN, A. (1914). Storage to be provided in impounding reservoirs for municipal water supply. *Trans. Amer. Soc. Civ. Engrs*, **77**, 1539–669.

HEMMINGSEN, A. M. (1933). The accuracy of insulin assay on white mice. *Quart. J. Pharm.* **6**, 39–80 and 187–217.

IRWIN, J. O. (1937). Statistical method applied to biological assay. *J. R. Statist. Soc. Suppl.* **4**, 1–60.

IRWIN, J. O. (1943). On the calculation of the error of biological assays. *J. Hyg., Camb.*, **43**, 121–8.

IRWIN, J. O. and CHEESEMAN, E. A. (1939a). On an approximate method of determining the median effective dose and its error in the case of a quantal response. *J. Hyg., Camb.*, **39**, 574–80.

IRWIN, J. O. and CHEESEMAN, E. A. (1939b). On the maximum likelihood method of determining dosage-response curves and approximations to the median-effective dose, in cases of a quantal response. *J. R. Statist. Soc. Suppl.* **6**, 174–85.

KÄRBER, G. (1931). Beitrag zur kollektiven Behandlung pharmakologischer Reihenversuche. *Arch. exp. Path. Pharmak.* **162**, 480–7.

KROGH, A. and HEMMINGSEN, A. M. (1928). The assay of insulin on rabbits and mice. *Biol. Medd., Kbh.*, **7**, 6.

LAWLEY, D. N. (1943). On problems connected with item selection and test construction. *Proc. Roy. Soc. Edinb.* A, **61**, 273–87.

LAWLEY, D. N. (1944). The factorial analysis of multiple item tests. *Proc. Roy. Soc. Edinb.* A, **62**, 74–82.

LE PELLEY, R. H. and SULLIVAN, W. N. (1936). Toxicity of rotenone and pyrethrins, alone and in combination. *J. Econ. Ent.* **29**, 791–6.

LITCHFIELD, J. T. and FERTIG, J. W. (1941). On a graphic solution of the dosage-effect curve. *Johns Hopk. Hosp. Bull.* **69**, 276–86.

McCALLAN, S. E. A. (1943). Empirical probit weights for dosage-response curves of greenhouse tomato foliage diseases. *Contr. Boyce Thompson Inst.* **13**, 177–84.

McLEOD, W. S. (1944). Further refinements of a technique for testing contact insecticides. *Canad. J. Res.* D, **22**, 87–104.

MARTIN, J. T. (1940). The problem of the evaluation of rotenone-containing plants. V. The relative toxicities of different species of derris. *Ann. Appl. Biol.* **27**, 274–94.

MARTIN, J. T. (1942). The problem of the evaluation of rotenone-containing plants. VI. The toxicity of l-elliptone and of poisons applied jointly, with further observations on the rotenone equivalent method of assessing the toxicity of derris root. *Ann. Appl. Biol.* **29**, 69–81.

MARTIN, J. T. (1943). The preparation of a standard pyrethrum extract in heavy mineral oil, with observations on the relative toxicities of the pyrethrins in oil and aqueous media. *Ann. Appl. Biol.* **30**, 293–300.

MATHER, K. (1943). *Statistical Analysis in Biology*. London: Methuen and Co., Ltd.

MILLER, L. C., BLISS, C. I. and BRAUN, H. A. (1939). The assay of digitalis. I. Criteria for evaluating various methods using frogs. *J. Amer. Pharm. Ass.* **28**, 644–57.

MONTGOMERY, H. B. S. and SHAW, H. (1943). Behaviour of thiuram sulphides, etc., in spore germination tests. *Nature, Lond.*, **151**, 333.

MOORE, W. and BLISS, C. I. (1942). A method for determining insecticidal effectiveness using *Aphis rumicis* and certain organic compounds. *J. Econ. Ent.* **35**, 544–53.

MÜLLER, G. E. (1879). Ueber die Maassbestimmungen des Ortsinnes der Haut mittels der Methode der richtigen und falschen Fälle. *Pflüg. Arch. ges. Physiol.* **19**, 191–235.

MURRAY, C. A. (1937). A statistical analysis of fly mortality data. *Soap*, **13**, no. 8, 89–105.

MURRAY, C. A. (1938). Dosage-mortality in the Peet-Grady method. *Soap*, **14**, no. 2, 99–103 and 123–5.

O'KANE, W. C., WESTGATE, W. A., GLOVER, L. C. and LOWRY, P. R. (1930). Studies of contact insecticides. I. *Tech. Bull. N. H. Agric. Exp. Sta.* no. 39.

O'KANE, W. C., WESTGATE, W. A. and GLOVER, L. C. (1934). Studies of contact insecticides. VII. *Tech. Bull. N. H. Agric. Exp. Sta.* no. 58.

OSTWALD, W. and DERNOSCHEK, A. (1910). Über die Beziehungen zwischen Adsorption und Giftigkeit. *Kolloidzschr.* **6**, 297–307.

PARKER-RHODES, A. F. (1941). Studies on the mechanism of fungicidal action. I. Preliminary investigation of nickel, copper, silver, and mercury. *Ann. Appl. Biol.* **28**, 389–405.

PARKER-RHODES, A. F. (1942a). Studies on the mechanism of fungicidal action. II. Elements of the theory of variability. *Ann. Appl. Biol.* **29**, 126–35.

PARKER-RHODES, A. F. (1942b). Studies on the mechanism of fungicidal action. III. Sulphur. *Ann. Appl. Biol.* **29**, 136–43.

PARKER-RHODES, A. F. (1942c). Studies on the mechanism of fungicidal action. IV. Mercury. *Ann. Appl. Biol.* **29**, 404–11.

PARKER-RHODES, A. F. (1943a). Studies in the mechanism of fungicidal action. V. Non-metallic and sodium dithiocarbamic acid derivatives. *Ann. Appl. Biol.* **30**, 170–8.

PARKER-RHODES, A. F. (1943b). Studies in the mechanism of fungicidal action. VI. Water. *Ann. Appl. Biol.* **30**, 372–9.

PETERS, G. and GANTER, W. (1935). Zur Frage der Abtötung des Kornkäfers mit Blausäure. *Z. angew. Ent.* **21**, 547–59.

RICHARDS, F. J. (1941). The diagrammatic representation of the results of physiological and other experiments designed factorially. *Ann. Bot., Lond.*, N.S. **5**, 249–62.

SHACKELL, L. F. (1923). Studies in protoplasm poisoning. I. Phenols. *J. Gen. Physiol.* **5**, 783–805.

STRAND, A. L. (1930). Measuring the toxicity of insect fumigants. *Industr. Engng Chem.* (Anal. ed.), **2**, 4–8.

TATTERSFIELD, F., GIMINGHAM, C. T. and MORRIS, H. M. (1925). Studies on contact insecticides. I. Introduction and methods. *Ann. Appl. Biol.* **12**, 60–5.

TATTERSFIELD, F. and MARTIN, J. T. (1935). The problem of the evaluation of rotenone-containing plants. I. *Derris elliptica* and *Derris malaccensis.* *Ann. Appl. Biol.* **22**, 578–605.

TATTERSFIELD, F. and MARTIN, J. T. (1938). The problem of the evaluation of rotenone-containing plants. IV. The toxicity to *Aphis rumicis* of certain products isolated from derris root. *Ann. Appl. Biol.* **25**, 411–29.

TATTERSFIELD, F. and MORRIS, H. M. (1924). An apparatus for testing the toxic values of contact insecticides under controlled conditions. *Bull. Ent. Res.* **14**, 223–33.

TATTERSFIELD, F. and POTTER, C. (1943). Biological methods of determining the insecticidal values of pyrethrum preparations (particularly extracts in heavy oil). *Ann. Appl. Biol.* **30**, 259–79.

THOMSON, G. H. (1914). The accuracy of the $\Phi(\gamma)$ process. *Brit. J. Psychol.* **7**, 44–55.

THOMSON, G. H. (1919). The criterion of goodness of fit of psychophysical curves. *Biometrika*, **12**, 216–30.

TREVAN, J. W. (1927). The error of determination of toxicity. *Proc. Roy. Soc.* B, **101**, 483–514.

URBAN, F. M. (1909). Die psychophysichen Massmethoden als Grundlagen empirischer Messungen. *Arch. ges. Psychol.* **15**, 261–355.

URBAN, F. M. (1910). Die psychophysichen Massmethoden als Grundlagen empirischer Messungen (continued). *Arch. ges. Psychol.* **16**, 168–227.

WHIPPLE, G. C. (1916). The element of chance in sanitation. *J. Franklin Inst.* **182**, 37–59 and 205–27.

WHITLOCK, J. H. and BLISS, C. I. (1943). A bio-assay technique for antihelmintics. *J. Parasit.* **29**, 48–58.

WILSON, E. B. and WORCESTER, J. (1943 a). The determination of LD 50 and its sampling error in bio-assay. *Proc. Nat. Acad. Sci., Wash.*, **29**, 79–85.

WILSON, E. B. and WORCESTER, J. (1943 b). The determination of LD 50 and its sampling error in bio-assay. II. *Proc. Nat. Acad. Sci., Wash.*, **29**, 114–20.

WILSON, E. B. and WORCESTER, J. (1943 c). Bio-assay on a general curve. *Proc. Nat. Acad. Sci., Wash.*, **29**, 150–4.

WILSON, E. B. and WORCESTER, J. (1943 d). The determination of LD 50 and its sampling error in bio-assay. III. *Proc. Nat. Acad. Sci., Wash.*, **29**, 257–62.

WOOD, E. C. (1944 a). Mathematics of biological assay. *Nature, Lond.*, **153**, 84–5.

WOOD, E. C. (1944 b). Mathematics of biological assay. *Nature, Lond.*, **153**, 680–1.

WORCESTER, J. and WILSON, E. B. (1943). A table determining LD 50 or the 50 % end-point. *Proc. Nat. Acad. Sci., Wash.*, **29**, 207–12.

WRIGHT, S. (1926). A frequency curve adapted to variation in percentage occurrence. *J. Amer. Statist. Ass.* **21**, 162–78.

YATES, F. (1937 a). The design and analysis of factorial experiments. *Tech. Commun. Bur. Soil Sci., Harpenden*, no. 35.

YATES, F. (1937 b). Incomplete randomized blocks. *Ann. Eugen., Lond.*, **7**, 121–40.

YATES, F. (1940). The recovery of inter-block information in balanced incomplete block designs. *Ann. Eugen., Lond.*, **10**, 317–25.

TABLE I. Transformation of Percentages to Probits

%	0·0	0·1	0·2	0·3	0·4	0·5	0·6	0·7	0·8	0·9	1	2	3	4	5
0	—	1·9098	2·1218	2·2522	2·3479	2·4242	2·4879	2·5427	2·5911	2·6344	For more detail see values for 95–100				
1	2·6737	2·7096	2·7429	2·7738	2·8027	2·8299	2·8556	2·8799	2·9031	2·9251					
2	2·9463	2·9665	2·9859	3·0046	3·0226	3·0400	3·0569	3·0732	3·0890	3·1043					
3	3·1192	3·1337	3·1478	3·1616	3·1750	3·1881	3·2009	3·2134	3·2256	3·2376					
4	3·2493	3·2608	3·2721	3·2831	3·2940	3·3046	3·3151	3·3253	3·3354	3·3454					
5	3·3551	3·3648	3·3742	3·3836	3·3928	3·4018	3·4107	3·4195	3·4282	3·4368	9	18	27	36	45
6	3·4452	3·4536	3·4618	3·4699	3·4780	3·4859	3·4937	3·5015	3·5091	3·5167	8	16	24	32	40
7	3·5242	3·5316	3·5389	3·5462	3·5534	3·5605	3·5675	3·5745	3·5813	3·5882	7	14	21	28	36
8	3·5949	3·6016	3·6083	3·6148	3·6213	3·6278	3·6342	3·6405	3·6468	3·6531	6	13	19	26	32
9	3·6592	3·6654	3·6715	3·6775	3·6835	3·6894	3·6953	3·7012	3·7070	3·7127	6	12	18	24	30
10	3·7184	3·7241	3·7298	3·7354	3·7409	3·7464	3·7519	3·7574	3·7628	3·7681	6	11	17	22	28
11	3·7735	3·7788	3·7840	3·7893	3·7945	3·7996	3·8048	3·8099	3·8150	3·8200	5	10	16	21	26
12	3·8250	3·8300	3·8350	3·8399	3·8448	3·8497	3·8545	3·8593	3·8641	3·8689	5	10	15	20	24
13	3·8736	3·8783	3·8830	3·8877	3·8923	3·8969	3·9015	3·9061	3·9107	3·9152	5	9	14	18	23
14	3·9197	3·9242	3·9286	3·9331	3·9375	3·9419	3·9463	3·9506	3·9550	3·9593	4	9	13	18	22
15	3·9636	3·9678	3·9721	3·9763	3·9806	3·9848	3·9890	3·9931	3·9973	4·0014	4	8	13	17	21
16	4·0055	4·0096	4·0137	4·0178	4·0218	4·0259	4·0299	4·0339	4·0379	4·0419	4	8	12	16	20
17	4·0458	4·0498	4·0537	4·0576	4·0615	4·0654	4·0693	4·0731	4·0770	4·0808	4	8	12	16	19
18	4·0846	4·0884	4·0922	4·0960	4·0998	4·1035	4·1073	4·1110	4·1147	4·1184	4	8	11	15	19
19	4·1221	4·1258	4·1295	4·1331	4·1367	4·1404	4·1440	4·1476	4·1512	4·1548	4	7	11	15	18
20	4·1584	4·1619	4·1655	4·1690	4·1726	4·1761	4·1796	4·1831	4·1866	4·1901	3	7	11	14	18
21	4·1936	4·1970	4·2005	4·2039	4·2074	4·2108	4·2142	4·2176	4·2210	4·2244	3	7	10	14	17
22	4·2278	4·2312	4·2345	4·2379	4·2412	4·2446	4·2479	4·2512	4·2546	4·2579	3	7	10	13	17
23	4·2612	4·2644	4·2677	4·2710	4·2743	4·2775	4·2808	4·2840	4·2872	4·2905	3	7	10	13	16
24	4·2937	4·2969	4·3001	4·3033	4·3065	4·3097	4·3129	4·3160	4·3192	4·3224	3	6	10	13	16
25	4·3255	4·3287	4·3318	4·3349	4·3380	4·3412	4·3443	4·3474	4·3505	4·3536	3	6	9	12	16
26	4·3567	4·3597	4·3628	4·3659	4·3689	4·3720	4·3750	4·3781	4·3811	4·3842	3	6	9	12	15
27	4·3872	4·3902	4·3932	4·3962	4·3992	4·4022	4·4052	4·4082	4·4112	4·4142	3	6	9	12	15
28	4·4172	4·4201	4·4231	4·4260	4·4290	4·4319	4·4349	4·4378	4·4408	4·4437	3	6	9	12	15
29	4·4466	4·4495	4·4524	4·4554	4·4583	4·4612	4·4641	4·4670	4·4698	4·4727	3	6	9	12	14

No.	0	1	2	3	4	5	6	7	8	9					
30	4·5013	4·4985	4·4956	4·4928	4·4899	4·4871	4·4842	4·4813	4·4785	4·4756	14	11	9	6	3
31	4·5295	4·5267	4·5239	4·5211	4·5183	4·5155	4·5126	4·5098	4·5070	4·5041	14	11	8	6	3
32	4·5573	4·5546	4·5518	4·5490	4·5462	4·5435	4·5407	4·5379	4·5351	4·5323	14	11	8	6	3
33	4·5848	4·5821	4·5793	4·5766	4·5739	4·5711	4·5684	4·5656	4·5628	4·5601	14	11	8	5	3
34	4·6120	4·6093	4·6066	4·6039	4·6011	4·5984	4·5957	4·5930	4·5903	4·5875	14	11	8	5	3
35	4·6389	4·6362	4·6335	4·6308	4·6281	4·6255	4·6228	4·6201	4·6174	4·6147	13	11	8	5	3
36	4·6655	4·6628	4·6602	4·6575	4·6549	4·6522	4·6495	4·6469	4·6442	4·6415	13	11	8	5	3
37	4·6919	4·6893	4·6866	4·6840	4·6814	4·6787	4·6761	4·6734	4·6708	4·6681	13	11	8	5	3
38	4·7181	4·7155	4·7129	4·7102	4·7076	4·7050	4·7024	4·6998	4·6971	4·6945	13	10	8	5	3
39	4·7441	4·7415	4·7389	4·7363	4·7337	4·7311	4·7285	4·7259	4·7233	4·7207	13	10	8	5	3
40	4·7699	4·7673	4·7647	4·7622	4·7596	4·7570	4·7544	4·7518	4·7492	4·7467	13	10	8	5	3
41	4·7955	4·7930	4·7904	4·7879	4·7853	4·7827	4·7802	4·7776	4·7750	4·7725	13	10	8	5	3
42	4·8211	4·8185	4·8160	4·8134	4·8109	4·8083	4·8058	4·8032	4·8007	4·7981	13	10	8	5	3
43	4·8465	4·8440	4·8414	4·8389	4·8363	4·8338	4·8313	4·8287	4·8262	4·8236	13	10	8	5	3
44	4·8718	4·8693	4·8668	4·8642	4·8617	4·8592	4·8566	4·8541	4·8516	4·8490	13	10	8	5	3
45	4·8970	4·8945	4·8920	4·8895	4·8870	4·8844	4·8819	4·8794	4·8769	4·8743	13	10	8	5	3
46	4·9222	4·9197	4·9172	4·9147	4·9122	4·9096	4·9071	4·9046	4·9021	4·8996	13	10	8	5	3
47	4·9473	4·9448	4·9423	4·9398	4·9373	4·9348	4·9323	4·9298	4·9272	4·9247	13	10	8	5	3
48	4·9724	4·9699	4·9674	4·9649	4·9624	4·9599	4·9574	4·9549	4·9524	4·9498	13	10	8	5	3
49	4·9975	4·9950	4·9925	4·9900	4·9875	4·9850	4·9825	4·9799	4·9774	4·9749	13	10	8	5	3
50	5·0226	5·0201	5·0175	5·0150	5·0125	5·0100	5·0075	5·0050	5·0025	5·0000	13	10	8	5	3
51	5·0476	5·0451	5·0426	5·0401	5·0376	5·0351	5·0326	5·0301	5·0276	5·0251	13	10	8	5	3
52	5·0728	5·0702	5·0677	5·0652	5·0627	5·0602	5·0577	5·0552	5·0527	5·0502	13	10	8	5	3
53	5·0979	5·0954	5·0929	5·0904	5·0878	5·0853	5·0828	5·0803	5·0778	5·0753	13	10	8	5	3
54	5·1231	5·1206	5·1181	5·1156	5·1130	5·1105	5·1080	5·1055	5·1030	5·1004	13	10	8	5	3
55	5·1484	5·1459	5·1434	5·1408	5·1383	5·1358	5·1332	5·1307	5·1282	5·1257	13	10	8	5	3
56	5·1738	5·1713	5·1687	5·1662	5·1637	5·1611	5·1586	5·1560	5·1535	5·1510	13	10	8	5	3
57	5·1993	5·1968	5·1942	5·1917	5·1891	5·1866	5·1840	5·1815	5·1789	5·1764	13	10	8	5	3
58	5·2250	5·2224	5·2198	5·2173	5·2147	5·2121	5·2096	5·2070	5·2045	5·2019	13	10	8	5	3
59	5·2508	5·2482	5·2456	5·2430	5·2404	5·2378	5·2353	5·2327	5·2301	5·2275	13	10	8	5	3

TABLE I (cont.)

%	0·0	0·1	0·2	0·3	0·4	0·5	0·6	0·7	0·8	0·9	1	2	3	4	5
60	5·2533	5·2559	5·2585	5·2611	5·2637	5·2663	5·2689	5·2715	5·2741	5·2767	3	5	8	10	13
61	5·2793	5·2819	5·2845	5·2871	5·2898	5·2924	5·2950	5·2976	5·3002	5·3029	3	5	8	10	13
62	5·3055	5·3081	5·3107	5·3134	5·3160	5·3186	5·3213	5·3239	5·3266	5·3292	3	5	8	11	13
63	5·3319	5·3345	5·3372	5·3398	5·3425	5·3451	5·3478	5·3505	5·3531	5·3558	3	5	8	11	13
64	5·3585	5·3611	5·3638	5·3665	5·3692	5·3719	5·3745	5·3772	5·3799	5·3826	3	5	8	11	13
65	5·3853	5·3880	5·3907	5·3934	5·3961	5·3989	5·4016	5·4043	5·4070	5·4097	3	5	8	11	14
66	5·4125	5·4152	5·4179	5·4207	5·4234	5·4261	5·4289	5·4316	5·4344	5·4372	3	5	8	11	14
67	5·4399	5·4427	5·4454	5·4482	5·4510	5·4538	5·4565	5·4593	5·4621	5·4649	3	6	8	11	14
68	5·4677	5·4705	5·4733	5·4761	5·4789	5·4817	5·4845	5·4874	5·4902	5·4930	3	6	8	11	14
69	5·4959	5·4987	5·5015	5·5044	5·5072	5·5101	5·5129	5·5158	5·5187	5·5215	3	6	9	11	14
70	5·5244	5·5273	5·5302	5·5330	5·5359	5·5388	5·5417	5·5446	5·5476	5·5505	3	6	9	12	14
71	5·5534	5·5563	5·5592	5·5622	5·5651	5·5681	5·5710	5·5740	5·5769	5·5799	3	6	9	12	15
72	5·5828	5·5858	5·5888	5·5918	5·5948	5·5978	5·6008	5·6038	5·6068	5·6098	3	6	9	12	15
73	5·6128	5·6158	5·6189	5·6219	5·6250	5·6280	5·6311	5·6341	5·6372	5·6403	3	6	9	12	15
74	5·6433	5·6464	5·6495	5·6526	5·6557	5·6588	5·6620	5·6651	5·6682	5·6713	3	6	9	12	16
75	5·6745	5·6776	5·6808	5·6840	5·6871	5·6903	5·6935	5·6967	5·6999	5·7031	3	6	10	13	16
76	5·7063	5·7095	5·7128	5·7160	5·7192	5·7225	5·7257	5·7290	5·7323	5·7356	3	7	10	13	16
77	5·7388	5·7421	5·7454	5·7488	5·7521	5·7554	5·7588	5·7621	5·7655	5·7688	3	7	10	13	17
78	5·7722	5·7756	5·7790	5·7824	5·7858	5·7892	5·7926	5·7961	5·7995	5·8030	3	7	10	14	17
79	5·8064	5·8099	5·8134	5·8169	5·8204	5·8239	5·8274	5·8310	5·8345	5·8381	4	7	11	14	18
80	5·8416	5·8452	5·8488	5·8524	5·8560	5·8596	5·8633	5·8669	5·8705	5·8742	4	7	11	14	18
81	5·8779	5·8816	5·8853	5·8890	5·8927	5·8965	5·9002	5·9040	5·9078	5·9116	4	7	11	15	19
82	5·9154	5·9192	5·9230	5·9269	5·9307	5·9346	5·9385	5·9424	5·9463	5·9502	4	8	12	15	19
83	5·9542	5·9581	5·9621	5·9661	5·9701	5·9741	5·9782	5·9822	5·9863	5·9904	4	8	12	16	20
84	5·9945	5·9986	6·0027	6·0069	6·0110	6·0152	6·0194	6·0237	6·0279	6·0322	4	8	13	17	21
85	6·0364	6·0407	6·0450	6·0494	6·0537	6·0581	6·0625	6·0669	6·0714	6·0758	4	9	13	18	22
86	6·0803	6·0848	6·0893	6·0939	6·0985	6·1031	6·1077	6·1123	6·1170	6·1217	5	9	14	18	23
87	6·1264	6·1311	6·1359	6·1407	6·1455	6·1503	6·1552	6·1601	6·1650	6·1700	5	10	15	19	24
88	6·1750	6·1800	6·1850	6·1901	6·1952	6·2004	6·2055	6·2107	6·2160	6·2212	5	10	15	21	26
89	6·2265	6·2319	6·2372	6·2426	6·2481	6·2536	6·2591	6·2646	6·2702	6·2759	5	11	16	22	27

%	0·00	0·01	0·02	0·03	0·04	0·05	0·06	0·07	0·08	0·09	1	2	3	4	5
90	6·2816	6·2873	6·2930	6·2988	6·3047	6·3106	6·3165	6·3225	6·3285	6·3346	6	12	18	24	29
91	6·3408	6·3469	6·3532	6·3595	6·3658	6·3722	6·3787	6·3852	6·3917	6·3984	6	13	19	26	32
92	6·4051	6·4118	6·4187	6·4255	6·4325	6·4395	6·4466	6·4538	6·4611	6·4684	7	14	21	28	35
93	6·4758	6·4833	6·4909	6·4985	6·5063	6·5141	6·5220	6·5301	6·5382	6·5464	8	16	24	31	39
94	6·5548	6·5632	6·5718	6·5805	6·5893	6·5982	6·6072	6·6164	6·6258	6·6352	9	18	27	36	45
95	6·6449	6·6546	6·6646	6·6747	6·6849	6·6954	6·7060	6·7169	6·7279	6·7392					
	97	100	101	102	105	106	109	110	113	115					
96	6·7507	6·7624	6·7744	6·7866	6·7991	6·8119	6·8250	6·8384	6·8522	6·8663					
	117	120	122	125	128	131	134	138	141	145					
97	6·8808	6·8957	6·9110	6·9268	6·9431	6·9600	6·9774	6·9954	7·0141	7·0335					
	149	153	158	163	169	174	180	187	194	202					

%	0·00	0·01	0·02	0·03	0·04	0·05	0·06	0·07	0·08	0·09	1	2	3	4	5
98·0	7·0537	7·0558	7·0579	7·0600	7·0621	7·0642	7·0663	7·0684	7·0706	7·0727	2	4	6	8	11
98·1	7·0749	7·0770	7·0792	7·0814	7·0836	7·0858	7·0880	7·0902	7·0924	7·0947	2	4	7	9	11
98·2	7·0969	7·0992	7·1015	7·1038	7·1061	7·1084	7·1107	7·1130	7·1154	7·1177	2	5	7	9	12
98·3	7·1201	7·1224	7·1248	7·1272	7·1297	7·1321	7·1345	7·1370	7·1394	7·1419	2	5	7	10	12
98·4	7·1444	7·1469	7·1494	7·1520	7·1545	7·1571	7·1596	7·1622	7·1648	7·1675	3	5	8	10	13
98·5	7·1701	7·1727	7·1754	7·1781	7·1808	7·1835	7·1862	7·1890	7·1917	7·1945	3	5	8	11	14
98·6	7·1973	7·2001	7·2029	7·2058	7·2086	7·2115	7·2144	7·2173	7·2203	7·2232	3	6	9	12	14
98·7	7·2262	7·2292	7·2322	7·2353	7·2383	7·2414	7·2445	7·2476	7·2508	7·2539	3	6	9	12	15
98·8	7·2571	7·2603	7·2636	7·2668	7·2701	7·2734	7·2768	7·2801	7·2835	7·2869	3	7	10	13	17
98·9	7·2904	7·2938	7·2973	7·3009	7·3044	7·3080	7·3116	7·3152	7·3189	7·3226	4	7	11	14	18
99·0	7·3263	7·3301	7·3339	7·3378	7·3416	7·3455	7·3495	7·3535	7·3575	7·3615	4	8	12	16	20
99·1	7·3656	7·3698	7·3739	7·3781	7·3824	7·3867	7·3911	7·3954	7·3999	7·4044	4	9	13	17	22
99·2	7·4089	7·4135	7·4181	7·4228	7·4276	7·4324	7·4372	7·4422	7·4471	7·4522	5	10	14	19	24
99·3	7·4573	7·4624	7·4677	7·4730	7·4783	7·4838	7·4893	7·4949	7·5006	7·5063	5	11	16	22	27
99·4	7·5121	7·5181	7·5241	7·5302	7·5364	7·5427	7·5491	7·5556	7·5622	7·5690	6	13	19	25	32
99·5	7·5758	7·5828	7·5899	7·5972	7·6045	7·6121	7·6197	7·6276	7·6356	7·6437					
99·6	7·6521	7·6606	7·6693	7·6783	7·6874	7·6968	7·7065	7·7164	7·7266	7·7370					
99·7	7·7478	7·7589	7·7703	7·7822	7·7944	7·8070	7·8202	7·8338	7·8480	7·8627					
99·8	7·8782	7·8943	7·9112	7·9290	7·9478	7·9677	7·9889	8·0115	8·0357	8·0618					
99·9	8·0902	8·1214	8·1559	8·1947	8·2389	8·2905	8·3528	8·4316	8·5401	8·7190					

I am indebted to Professor R. A. Fisher and Dr F. Yates, and also to Messrs Oliver and Boyd, Ltd. of Edinburgh, for permission to reprint Table I from Table IX of their book *Statistical Tables for Biological, Agricultural and Medical Research*.

TABLE II. The Weighting Coefficient and Q/Z

Y	Q/Z	Percentage natural mortality, C										
		0	1	2	3	4	5	6	7	8	9	10
1·1	5034	·00082	—	—	—	—	—	—	—	—	—	—
1·2	3425	·00118	·00001	—	—	—	—	—	—	—	—	—
1·3	2354	·00167	·00002	·00001	·00001	—	—	—	—	—	—	—
1·4	1634	·00235	·00004	·00002	·00001	·00001	—	·00001	—	—	—	—
1·5	1146	·00327	·00007	·00004	·00002	·00002	·00001	·00001	·00001	·00001	·00001	·00001
1·6	811·2	·00451	·00015	·00007	·00005	·00004	·00003	·00002	·00002	·00002	·00002	·00001
1·7	580·2	·00614	·00028	·00014	·00009	·00007	·00006	·00005	·00004	·00003	·00003	·00003
1·8	419·1	·00828	·00053	·00027	·00018	·00013	·00011	·00009	·00007	·00006	·00006	·00005
1·9	305·8	·01105	·00097	·00050	·00034	·00025	·00020	·00017	·00014	·00012	·00011	·00010
2·0	225·3	·01457	·00172	·00090	·00061	·00046	·00036	·00030	·00026	·00022	·00020	·00017
2·1	167·69	·01903	·00297	·00159	·00108	·00082	·00065	·00054	·00046	·00040	·00035	·00031
2·2	126·02	·02458	·00496	·00274	·00188	·00142	·00114	·00095	·00081	·00070	·00062	·00055
2·3	95·63	·03143	·00803	·00456	·00317	·00241	·00194	·00162	·00138	·00121	·00106	·00095
2·4	73·28	·03977	·01256	·00739	·00521	·00400	·00324	·00271	·00232	·00202	·00179	·00160
2·5	56·70	·04979	·01895	·01161	·00832	·00646	·00525	·00441	·00379	·00332	·00294	·00264
2·6	44·288	·06168	·02763	·01768	·01292	·01014	·00831	·00702	·00606	·00531	·00472	·00424
2·7	34·923	·07563	·03895	·02605	·01947	·01548	·01280	·01088	·00943	·00830	·00740	·00666
2·8	27·797	·09179	·05316	·03719	·02847	·02297	·01918	·01642	·01431	·01265	·01131	·01021
2·9	22·330	·11026	·07044	·05147	·04037	·03309	·02794	·02411	·02115	·01879	·01687	·01527
3·0	18·101	·13112	·09080	·06912	·05557	·04631	·03957	·03445	·03043	·02719	·02452	·02228
3·1	14·802	·15436	·11419	·09023	·07432	·06298	·05449	·04790	·04263	·03832	·03473	·03170
3·2	12·211	·17994	·14046	·11476	·09670	·08332	·07300	·06481	·05814	·05261	·04795	·04397
3·3	10·159	·20773	·16935	·14249	·12263	·10736	·09525	·08541	·07726	·07039	·06453	·05947
3·4	8·521	·23753	·20056	·17308	·15184	·13494	·12116	·10973	·10008	·09182	·08469	·07846
3·5	7·205	·26907	·23373·	·20611	·18392	·16571	·15050	·13760	·12652	·11690	·10848	·10103

x												
3·6	·12711	·13575	·14541	·15631	·16867	·18283	·19921	·21836	·24107	·26842	·30199	6·1394
3·7	·15639	·16614	·17694	·18896	·20242	·21759	·23482	·25456	·27741	·30415	·33589	5·2705
3·8	·18840	·19915	·21092	·22387	·23819	·25409	·27187	·29186	·31453	·34043	·37031	4·4571
3·9	·22250	·23409	·24665	·26031	·27524	·29161	·30964	·32960	·35181	·37669	·40474	3·9676
4·0	·25797	·27020	·28334	·29749	·31279	·32937	·34739	·36707	·38864	·41237	·43863	3·4770
4·1	·29397	·30666	·32017	·33460	·35005	·36661	·38441	·40362	·42438	·44691	·47144	3·0665
4·2	·32969	·34264	·35634	·37085	·38623	·40259	·42000	·43858	·45844	·47973	·50260	2·7206
4·3	·36430	·37735	·39105	·40546	·42063	·43662	·45350	·47134	·49024	·51029	·53159	2·4276
4·4	·39702	·41002	·42357	·43774	·45255	·46805	·48430	·50134	·51924	·53806	·55788	2·1780
4·5	·42716	·43996	·45325	·46705	·48140	·49633	·51187	·52806	·54495	·56257	·58099	1·9640
4·6	·45409	·46659	·47951	·49286	·50666	·52095	·53574	·55106	·56694	·58341	·60052	1·7797
4·7	·47729	·48941	·50187	·51470	·52790	·54150	·55551	·56996	·58485	·60022	·61609	1·6202
4·8	·49635	·50801	·51996	·53221	·54478	·55766	·57089	·58446	·59840	·61271	·62742	1·4814
4·9	·51094	·52210	·53350	·54514	·55704	·56921	·58164	·59436	·60737	·62069	·63431	1·3599
5·0	·52087	·53149	·54230	·55332	·56455	·57599	·58765	·59953	·61165	·62401	·63662	1·2533
5·1	·52604	·53609	·54631	·55669	·56724	·57796	·58886	·59994	·61120	·62266	·63431	1·1593
5·2	·52644	·53592	·54553	·55527	·56515	·57516	·58532	·59562	·60607	·61667	·62742	1·0759
5·3	·52219	·53108	·54008	·54919	·55841	·56773	·57717	·58672	·59639	·60618	·61609	1·0018
5·4	·51347	·52178	·53018	·53866	·54722	·55588	·56462	·57346	·58238	·59140	·60052	0·9357
5·5	·50056	·50829	·51609	·52396	·53189	·53990	·54797	·55612	·56434	·57263	·58099	0·8764
5·6	·48380	·49097	·49818	·50545	·51278	·52015	·52759	·53507	·54262	·55022	·55788	0·8230
5·7	·46363	·47024	·47688	·48357	·49030	·49708	·50389	·51075	·51765	·52460	·53159	0·7749
5·8	·44050	·44657	·45266	·45879	·46495	·47114	·47737	·48363	·48992	·49624	·50260	0·7313
5·9	·41493	·42047	·42603	·43162	·43723	·44287	·44853	·45422	·45993	·46567	·47144	0·6917
6·0	·38746	·39249	·39754	·40261	·40770	·41281	·41793	·42308	·42824	·43343	·43863	0·6557
6·1	·35863	·36318	·36774	·37231	·37690	·38150	·38612	·39075	·39540	·40006	·40474	0·6227
6·2	·32900	·33308	·33718	·34128	·34540	·34952	·35366	·35781	·36196	·36613	·37031	0·5926
6·3	·29910	·30274	·30640	·31006	·31372	·31740	·32108	·32477	·32847	·33218	·33589	0·5649
6·4	·26942	·27266	·27589	·27913	·28238	·28564	·28890	·29216	·29543	·29871	·30199	0·5394
6·5	·24044	·24329	·24613	·24899	·25184	·25470	·25757	·26044	·26331	·26619	·26907	0·5158

TABLE II (cont.)

Percentage natural mortality, C

Y	Q/Z	0	1	2	3	4	5	6	7	8	9	10
6·6	0·4940	·23753	·23502	·23251	·23001	·22751	·22501	·22251	·22001	·21752	·21503	·21255
6·7	0·4739	·20773	·20556	·20339	·20122	·19905	·19689	·19473	·19256	·19041	·18825	·18609
6·8	0·4551	·17994	·17808	·17621	·17435	·17249	·17063	·16877	·16691	·16506	·16320	·16135
6·9	0·4376	·15436	·15277	·15118	·14960	·14801	·14643	·14484	·14326	·14168	·14010	·13852
7·0	0·4214	·13112	·12977	·12843	·12709	·12575	·12442	·12308	·12174	·12040	·11907	·11773
7·1	0·4062	·11026	·10914	·10802	·10689	·10577	·10465	·10353	·10241	·10129	·10017	·09905
7·2	0·3919	·09179	·09086	·08993	·08900	·08807	·08714	·08621	·08528	·08435	·08342	·08249
7·3	0·3786	·07564	·07487	·07411	·07334	·07258	·07181	·07105	·07029	·06952	·06876	·06800
7·4	0·3661	·06168	·06106	·06044	·05982	·05920	·05858	·05795	·05733	·05671	·05609	·05547
7·5	0·3543	·04979	·04929	·04879	·04828	·04778	·04728	·04678	·04628	·04578	·04528	·04478
7·6	0·3432	·03977	·03937	·03897	·03857	·03817	·03777	·03737	·03697	·03657	·03617	·03577
7·7	0·3327	·03143	·03112	·03080	·03048	·03017	·02985	·02954	·02922	·02891	·02859	·02828
7·8	0·3228	·02458	·02434	·02409	·02385	·02360	·02335	·02311	·02286	·02261	·02237	·02212
7·9	0·3134	·01903	·01883	·01864	·01845	·01826	·01807	·01788	·01769	·01750	·01731	·01712
8·0	0·3046	·01457	·01442	·01428	·01413	·01399	·01384	·01369	·01355	·01340	·01326	·01311
8·1	0·2962	·01104	·01093	·01082	·01071	·01060	·01049	·01038	·01027	·01016	·01005	·00993
8·2	0·2882	·00828	·00819	·00811	·00803	·00795	·00786	·00778	·00770	·00762	·00753	·00745
8·3	0·2806	·00614	·00608	·00602	·00596	·00590	·00583	·00577	·00571	·00565	·00559	·00553
8·4	0·2734	·00451	·00446	·00442	·00437	·00433	·00428	·00424	·00419	·00415	·00410	·00406
8·5	0·2666	·00327	·00324	·00321	·00318	·00314	·00311	·00308	·00305	·00301	·00298	·00295
8·6	0·2600	·00235	·00233	·00231	·00228	·00226	·00224	·00221	·00219	·00217	·00214	·00212
8·7	0·2538	·00167	·00166	·00164	·00162	·00161	·00159	·00157	·00156	·00154	·00152	·00150
8·8	0·2478	·00118	·00117	·00116	·00114	·00113	·00112	·00111	·00110	·00108	·00107	·00106
8·9	0·2421	·00082	·00081	·00080	·00080	·00079	·00078	·00077	·00076	·00076	·00075	·00074
9·0	0·2367	·00056	·00056	·00055	·00055	·00054	·00054	·00053	·00053	·00052	·00051	·00051

Y	Q/Z	11	12	13	14	15	16	17	18	19	20
1·1	5034	—	—	—	—	—	—	—	—	—	—
1·2	3425	—	—	—	—	—	—	—	—	—	—
1·3	2354	—	—	—	—	—	—	—	—	—	—
1·4	1634	·00001	—	—	—	—	—	—	—	—	—
1·5	1146	·00001	·00001	·00001	—	—	—	—	—	—	—
1·6	811·2	·00001	·00001	·00001	·00001	·00001	·00001	·00001	·00001	·00001	·00001
1·7	580·2	·00002	·00002	·00002	·00002	·00002	·00002	·00001	·00001	·00001	·00001
1·8	419·1	·00005	·00004	·00004	·00003	·00003	·00003	·00003	·00003	·00002	·00002
1·9	305·8	·00009	·00008	·00007	·00007	·00006	·00006	·00005	·00005	·00005	·00004
2·0	225·3	·00016	·00014	·00013	·00012	·00011	·00010	·00010	·00009	·00008	·00008
2·1	167·69	·00028	·00026	·00023	·00022	·00020	·00018	·00017	·00016	·00015	·00014
2·2	126·02	·00050	·00045	·00041	·00038	·00035	·00033	·00030	·00028	·00026	·00025
2·3	95·63	·00086	·00078	·00071	·00066	·00061	·00056	·00052	·00049	·00046	·00043
2·4	73·28	·00145	·00131	·00120	·00111	·00102	·00095	·00088	·00083	·00077	·00073
2·5	56·70	·00238	·00217	·00199	·00183	·00169	·00157	·00147	·00137	·00128	·00121
2·6	44·288	·00384	·00350	·00321	·00296	·00274	·00255	·00237	·00222	·00208	·00196
2·7	34·923	·00604	·00551	·00506	·00467	·00433	·00403	·00376	·00352	·00331	·00311
2·8	27·797	·00928	·00849	·00781	·00722	·00670	·00624	·00583	·00547	·00514	·00484
2·9	22·330	·01392	·01277	·01177	·01090	·01014	·00945	·00885	·00830	·00780	·00735
3·0	18·101	·02038	·01875	·01732	·01608	·01497	·01399	·01311	·01231	·01159	·01094
3·1	14·802	·02910	·02685	·02488	·02315	·02160	·02022	·01898	·01786	·01684	·01590
3·2	12·211	·04053	·03753	·03488	·03254	·03044	·02856	·02686	·02531	·02390	·02261
3·3	10·159	·05505	·05117	·04772	·04465	·04188	·03939	·03712	·03506	·03317	·03143
3·4	8·521	·07297	·06809	·06374	·05982	·05628	·05307	·05014	·04745	·04498	·04271
3·5	7·205	·09441	·08848	·08313	·07829	·07389	·06987	·06618	·06278	·05965	·05674
3·6	6·1394	·11934	·11232	·10595	·10014	·09481	·08991	·08540	·08122	·07734	·07373
3·7	5·2705	·14753	·13945	·13205	·12625	·11898	·11318	·10780	·10279	·09812	·09376
3·8	4·5571	·17854	·16947	·16111	·15336	·14616	·13946	·13321	·12736	·12187	·11672
3·9	3·9676	·21179	·20185	·19260	·18398	·17591	·16836	·16127	·15460	·14831	·14237
4·0	3·4770	·24656	·23589	·22589	·21649	·20766	·19933	·19146	·18402	·17698	·17029

TABLE II (cont.)

Percentage natural mortality, C

Y	Q/Z	11	12	13	14	15	16	17	18	19	20
4·1	3·0665	·28204	·27081	·26020	·25017	·24068	·23168	·22314	·21501	·20728	·19991
4·2	2·7206	·31742	·30578	·29473	·28421	·27420	·26465	·25554	·24684	·23852	·23055
4·3	2·4276	·35186	·33998	·32864	·31779	·30740	·29744	·28789	·27873	·26992	·26145
4·4	2·1780	·38457	·37261	·36112	·35008	·33945	·32922	·31937	·30986	·30069	·29184
4·5	1·9640	·41482	·40292	·39142	·38033	·36960	·35922	·34919	·33947	·33006	·32094
4·6	1·7797	·44198	·43025	·41887	·40784	·39713	·38674	·37664	·36683	·35729	·34802
4·7	1·6202	·46551	·45405	·44289	·43202	·42144	·41113	·40109	·39129	·38174	·37242
4·8	1·4814	·48496	·47385	·46299	·46239	·44202	·43190	·42199	·41231	·40284	·39357
4·9	1·3599	·50001	·48931	·47883	·46855	·45849	·44862	·43894	·42945	·42015	·41102
5·0	1·2533	·51044	·50020	·49014	·48026	·47054	·46100	·45162	·44240	·43333	·42441
5·1	1·1593	·51614	·50639	·49680	·48735	·47804	·46887	·45984	·45094	·44217	·43354
5·2	1·0759	·51709	·50787	·49876	·48978	·48091	·47216	·46353	·45500	·44658	·43827
5·3	1·0018	·51340	·50471	·49612	·48762	·47923	·47092	·46271	·45459	·44657	·43863
5·4	0·9357	·50524	·49709	·48903	·48104	·47313	·46529	·45754	·44985	·44224	·43470
5·5	0·8764	·49289	·48529	·47775	·47028	·46286	·45551	·44822	·44099	·43382	·42671
5·6	0·8230	·47669	·46963	·46262	·45567	·44876	·44190	·43509	·42832	·42161	·41494
5·7	0·7749	·45706	·45054	·44406	·43761	·43120	·42484	·41851	·41222	·40597	·39975
5·8	0·7313	·43447	·42847	·42250	·41656	·41066	·40478	·39893	·39311	·38733	·38157
5·9	0·6917	·40942	·40393	·39846	·39302	·38761	·38221	·37684	·37149	·36617	·36087
6·0	0·6557	·38245	·37745	·37248	·36752	·36258	·35766	·35275	·34787	·34300	·33815
6·1	0·6227	·35410	·34958	·34508	·34059	·33611	·33165	·32720	·32276	·31834	·31393
6·2	0·5926	·32493	·32087	·31681	·31277	·30874	·30472	·30071	·29671	·29272	·28874
6·3	0·5649	·29546	·29183	·28821	·28460	·28099	·27739	·27380	·27022	·26664	·26308
6·4	0·5394	·26620	·26298	·25977	·25656	·25335	·25016	·24696	·24378	·24060	·23742
6·5	0·5158	·23760	·23476	·23193	·22910	·22628	·22346	·22064	·21783	·21502	·21222

6·6	0·4940	·21007	·20759	·20511	·20264	·20016	·19770	·19523	·19277	·19031	·18785
6·7	0·4739	·18394	·18179	·17964	·17749	·17535	·17320	·17106	·16892	·16679	·16465
6·8	0·4551	·15950	·15765	·15580	·15395	·15210	·15026	·14841	·14657	·14473	·14289
6·9	0·4376	·13694	·13536	·13378	·13220	·13063	·12905	·12748	·12591	·12433	·12276
7·0	0·4214	·11640	·11506	·11373	·11239	·11106	·10973	·10840	·10707	·10574	·10441
7·1	0·4062	·09794	·09682	·09570	·09458	·09347	·09235	·09123	·09012	·08900	·08789
7·2	0·3919	·08157	·08064	·07971	·07878	·07786	·07693	·07600	·07508	·07415	·07323
7·3	0·3786	·06724	·06647	·06571	·06495	·06419	·06342	·06266	·06190	·06114	·06038
7·4	0·3661	·05485	·05423	·05361	·05299	·05237	·05175	·05113	·05051	·04989	·04927
7·5	0·3543	·04428	·04378	·04328	·04278	·04228	·04178	·04128	·04078	·04028	·03978
7·6	0·3432	·03537	·03498	·03458	·03418	·03378	·03338	·03298	·03258	·03218	·03178
7·7	0·3327	·02796	·02765	·02733	·02702	·02670	·02639	·02607	·02576	·02544	·02513
7·8	0·3228	·02187	·02163	·02138	·02114	·02089	·02064	·02040	·02015	·01990	·01966
7·9	0·3134	·01693	·01674	·01655	·01636	·01617	·01598	·01579	·01560	·01541	·01521
8·0	0·3046	·01297	·01282	·01267	·01253	·01238	·01224	·01209	·01194	·01180	·01165
8·1	0·2962	·00982	·00971	·00960	·00949	·00938	·00927	·00916	·00905	·00894	·00883
8·2	0·2882	·00737	·00728	·00720	·00712	·00704	·00695	·00687	·00679	·00670	·00662
8·3	0·2806	·00547	·00540	·00534	·00528	·00522	·00516	·00510	·00504	·00497	·00491
8·4	0·2734	·00401	·00397	·00392	·00388	·00383	·00379	·00374	·00370	·00365	·00361
8·5	0·2666	·00291	·00288	·00285	·00282	·00278	·00275	·00272	·00269	·00265	·00262
8·6	0·2600	·00209	·00207	·00205	·00202	·00200	·00198	·00195	·00193	·00191	·00188
8·7	0·2538	·00149	·00147	·00145	·00144	·00142	·00140	·00139	·00137	·00135	·00134
8·8	0·2478	·00105	·00104	·00103	·00101	·00100	·00099	·00098	·00097	·00096	·00094
8·9	0·2421	·00073	·00072	·00071	·00071	·00070	·00069	·00068	·00067	·00067	·00066
9·0	0·2367	·00050	·00050	·00049	·00049	·00048	·00047	·00047	·00046	·00046	·00045

TABLE II (cont.)

Percentage natural mortality, C

Y	Q/Z	21	22	23	24	25	26	27	28	29	30
1·1	5034	—	—	—	—	—	—	—	—	—	—
1·2	3425	—	—	—	—	—	—	—	—	—	—
1·3	2354	—	—	—	—	—	—	—	—	—	—
1·4	1634	—	—	—	—	—	—	—	—	—	—
1·5	1146	—	—	—	—	—	—	—	—	—	—
1·6	811·2	·00001	·00001	·00001	—	—	—	—	—	—	—
1·7	580·2	·00001	·00001	·00001	·00001	·00001	·00001	·00001	·00001	·00001	·00001
1·8	419·1	·00002	·00002	·00002	·00002	·00002	·00002	·00002	·00001	·00001	·00001
1·9	305·8	·00004	·00004	·00004	·00003	·00003	·00003	·00003	·00003	·00003	·00002
2·0	225·3	·00007	·00007	·00007	·00006	·00006	·00006	·00005	·00005	·00005	·00005
2·1	167·69	·00013	·00013	·00012	·00011	·00011	·00010	·00010	·00009	·00009	·00008
2·2	126·02	·00023	·00022	·00021	·00020	·00019	·00018	·00017	·00016	·00015	·00015
2·3	95·63	·00040	·00038	·00036	·00034	·00032	·00031	·00029	·00028	·00026	·00025
2·4	73·28	·00069	·00065	·00061	·00058	·00055	·00052	·00049	·00047	·00045	·00043
2·5	56·70	·00114	·00107	·00101	·00096	·00091	·00086	·00082	·00078	·00075	·00071
2·6	44·288	·00185	·00174	·00165	·00156	·00148	·00141	·00134	·00127	·00121	·00116
2·7	34·923	·00293	·00277	·00262	·00248	·00236	·00224	·00213	·00203	·00194	·00185
2·8	27·797	·00456	·00431	·00408	·00387	·00368	·00349	·00333	·00317	·00302	·00288
2·9	22·330	·00694	·00657	·00622	·00590	·00561	·00533	·00508	·00484	·00462	·00441
3·0	18·101	·01034	·00979	·00928	·00881	·00838	·00797	·00760	·00725	·00692	·00661
3·1	14·802	·01505	·01426	·01354	·01287	·01224	·01166	·01112	·01061	·01014	·00969
3·2	12·211	·02143	·02033	·01932	·01838	·01751	·01669	·01593	·01522	·01455	·01392
3·3	10·159	·02983	·02834	·02697	·02569	·02450	·02338	·02234	·02136	·02044	·01957
3·4	8·521	·04060	·03864	·03682	·03512	·03354	·03205	·03065	·02934	·02810	·02693
3·5	7·205	·05404	·05153	·04918	·04698	·04492	·04299	·04117	·03945	·03782	·03629

·04788	·04985	·05193	·05412	·05644	·05889	·06150	·06427	·06722	·07037	6·1394	3·6
·06189	·06435	·06695	·06967	·07255	·07559	·07881	·08221	·08582	·08966	5·2705	3·7
·07838	·08139	·08455	·08787	·09136	·09503	·09890	·10298	·10730	·11187	4·5571	3·8
·09732	·10091	·10468	·10862	·11275	·11708	·12163	·12641	·13145	·13676	3·9676	3·9
·11851	·12271	·12710	·13167	·13645	·14145	·14668	·15216	·15791	·16394	3·4770	4·0
·14164	·14645	·15145	·15665	·16207	·16771	·17360	·17974	·18616	·19288	3·0665	4·1
·16626	·17166	·17725	·18304	·18906	·19531	·20180	·20855	·21559	·22291	2·7206	4·2
·19182	·19776	·20389	·21023	·21679	·22358	·23061	·23790	·24546	·25331	2·4276	4·3
·21769	·22411	·23072	·23753	·24456	·25181	·25930	·26704	·27503	·28329	2·1780	4·4
·24319	·25002	·25703	·26424	·27165	·27927	·28712	·29520	·30352	·31210	1·9640	4·5
·26764	·27479	·28212	·28963	·29734	·30524	·31335	·32167	·33022	·33900	1·7797	4·6
·29038	·29777	·30533	·31305	·32095	·32904	·33731	·34578	·35444	·36332	1·6202	4·7
·31082	·31836	·32605	·33390	·34190	·35007	·35841	·36693	·37562	·38450	1·4814	4·8
·32843	·33604	·34378	·35166	·35968	·36785	·37617	·38464	·39327	·40206	1·3599	4·9
·34279	·35039	·35810	·36593	·37389	·38197	·39019	·39853	·40702	·41564	1·2533	5·0
·35359	·36109	·36870	·37641	·38423	·39216	·40020	·40836	·41663	·42502	1·1593	5·1
·36062	·36796	·37540	·38292	·39054	·39825	·40606	·41396	·42196	·43007	1·0759	5·2
·36378	·37091	·37812	·38540	·39276	·40020	·40772	·41532	·42300	·43077	1·0018	5·3
·36309	·36996	·37689	·38388	·39094	·39807	·40526	·41252	·41984	·42724	0·9357	5·4
·35868	·36524	·37185	·37852	·38524	·39201	·39884	·40572	·41266	·41966	0·8764	5·5
·35075	·35697	·36324	·36954	·37590	·38229	·38873	·39521	·40174	·40832	0·8230	5·6
·33959	·34545	·35134	·35727	·36323	·36923	·37526	·38133	·38743	·39357	0·7749	5·7
·32557	·33104	·33655	·34207	·34763	·35322	·35883	·36447	·37014	·37584	0·7313	5·8
·30909	·31417	·31927	·32439	·32954	·33470	·33989	·34510	·35033	·35559	0·6917	5·9
·29060	·29528	·29997	·30469	·30941	·31416	·31892	·32370	·32850	·33332	0·6557	6·0
·27057	·27485	·27914	·28344	·28776	·29209	·29643	·30079	·30516	·30954	0·6227	6·1
·24949	·25337	·25726	·26116	·26507	·26899	·27292	·27686	·28081	·28477	0·5926	6·2
·22780	·23130	·23480	·23831	·24182	·24535	·24888	·25242	·25596	·25952	0·5649	6·3
·20597	·20909	·21221	·21535	·21848	·22163	·22477	·22793	·23109	·23425	0·5394	6·4
·18439	·18715	·18992	·19270	·19547	·19825	·20104	·20383	·20662	·20942	0·5158	6·5

TABLE II (cont.)

Percentage natural mortality, C

Y	Q/Z	21	22	23	24	25	26	27	28	29	30
6·6	0·4940	·18540	·18294	·18049	·17805	·17561	·17317	·17073	·16829	·16586	·16343
6·7	0·4739	·16252	·16039	·15826	·15613	·15401	·15188	·14976	·14764	·14552	·14341
6·8	0·4551	·14105	·13921	·13738	·13554	·13371	·13188	·13005	·12822	·12639	·12457
6·9	0·4376	·12119	·11962	·11805	·11649	·11492	·11336	·11179	·11023	·10866	·10710
7·0	0·4214	·10308	·10175	·10042	·09909	·09777	·09644	·09512	·09379	·09247	·09114
7·1	0·4062	·08677	·08566	·08455	·08343	·08232	·08121	·08010	·07899	·07787	·07676
7·2	0·3919	·07230	·07137	·07045	·06953	·06860	·06768	·06675	·06583	·06491	·06398
7·3	0·3786	·05962	·05886	·05809	·05733	·05657	·05581	·05505	·05429	·05353	·05277
7·4	0·3661	·04865	·04803	·04741	·04679	·04617	·04555	·04493	·04431	·04369	·04307
7·5	0·3543	·03928	·03878	·03828	·03778	·03728	·03678	·03628	·03578	·03528	·03479
7·6	0·3432	·03139	·03099	·03059	·03019	·02979	·02939	·02899	·02860	·02820	·02780
7·7	0·3327	·02481	·02450	·02418	·02387	·02355	·02324	·02292	·02261	·02229	·02198
7·8	0·3228	·01941	·01917	·01892	·01867	·01843	·01818	·01793	·01769	·01744	·01720
7·9	0·3134	·01502	·01483	·01464	·01445	·01426	·01407	·01388	·01369	·01350	·01331
8·0	0·3046	·01151	·01136	·01122	·01107	·01092	·01078	·01063	·01049	·01034	·01019
8·1	0·2962	·00872	·00861	·00850	·00839	·00828	·00817	·00806	·00795	·00784	·00773
8·2	0·2882	·00654	·00646	·00637	·00629	·00621	·00612	·00604	·00596	·00588	·00579
8·3	0·2806	·00485	·00479	·00473	·00467	·00461	·00454	·00448	·00442	·00436	·00430
8·4	0·2734	·00356	·00352	·00347	·00343	·00338	·00334	·00329	·00325	·00320	·00316
8·5	0·2666	·00259	·00255	·00252	·00249	·00246	·00242	·00239	·00236	·00232	·00229
8·6	0·2600	·00186	·00184	·00181	·00179	·00176	·00174	·00172	·00169	·00167	·00165
8·7	0·2538	·00132	·00130	·00129	·00127	·00125	·00124	·00122	·00120	·00119	·00117
8·8	0·2478	·00093	·00092	·00091	·00090	·00088	·00087	·00086	·00085	·00084	·00083
8·9	0·2421	·00065	·00064	·00063	·00062	·00062	·00061	·00060	·00059	·00058	·00057
9·0	0·2367	·00045	·00044	·00044	·00043	·00042	·00042	·00041	·00041	·00040	·00040

Y	Q/Z	31	32	33	34	35	36	37	38	39	40
1·1	5034	—	—	—	—	—	—	—	—	—	—
1·2	3425	—	—	—	—	—	—	—	—	—	—
1·3	2354	—	—	—	—	—	—	—	—	—	—
1·4	1634	—	—	—	—	—	—	—	—	—	—
1·5	1146	—	—	—	—	—	—	—	—	—	—
1·6	811·2	—	—	—	—	—	—	—	—	—	—
1·7	580·2	·00001	·00001	·00001	·00001	·00001	·00001	·00001	·00001	—	—
1·8	419·1	·00001	·00001	·00001	·00001	·00001	·00001	·00001	·00001	·00001	·00001
1·9	305·8	·00002	·00002	·00002	·00002	·00002	·00002	·00002	·00002	·00002	·00002
2·0	225·3	·00004	·00004	·00004	·00004	·00004	·00003	·00003	·00003	·00003	·00003
2·1	167·69	·00008	·00008	·00007	·00007	·00007	·00006	·00006	·00006	·00006	·00005
2·2	126·02	·00014	·00013	·00013	·00012	·00012	·00011	·00011	·00010	·00010	·00009
2·3	95·63	·00024	·00023	·00022	·00021	·00020	·00019	·00018	·00018	·00017	·00016
2·4	73·28	·00041	·00039	·00037	·00036	·00034	·00033	·00031	·00030	·00029	·00028
2·5	56·70	·00068	·00065	·00062	·00059	·00057	·00054	·00052	·00050	·00048	·00046
2·6	44·288	·00111	·00106	·00101	·00097	·00092	·00089	·00085	·00081	·00078	·00075
2·7	34·923	·00176	·00169	·00161	·00154	·00148	·00142	·00136	·00130	·00125	·00120
2·8	27·797	·00276	·00263	·00252	·00241	·00231	·00221	·00212	·00204	·00195	·00188
2·9	22·330	·00422	·00403	·00386	·00370	·00354	·00339	·00325	·00312	·00300	·00288
3·0	18·101	·00632	·00605	·00579	·00555	·00532	·00510	·00489	·00469	·00451	·00433
3·1	14·802	·00927	·00888	·00850	·00815	·00782	·00750	·00720	·00691	·00664	·00637
3·2	12·211	·01333	·01276	·01223	·01173	·01126	·01080	·01037	·00996	·00957	·00920
3·3	10·159	·01875	·01797	·01724	·01654	·01588	·01525	·01465	·01408	·01354	·01302
3·4	8·521	·02582	·02478	·02378	·02284	·02194	·02109	·02027	·01949	·01875	·01804
3·5	7·205	·03483	·03345	·03214	·03089	·02970	·02856	·02748	·02645	·02546	·02451

Table II (cont.)

Percentage natural mortality, C

Y	Q/Z	31	32	33	34	35	36	37	38	39	40
3·6	6·1394	·04601	·04423	·04254	·04093	·03938	·03791	·03651	·03516	·03387	·03263
3·7	5·2705	·05954	·05731	·05517	·05313	·05118	·04932	·04753	·04581	·04417	·04259
3·8	4·5571	·07551	·07276	·07013	·06761	·06520	·06289	·06067	·05853	·05648	·05451
3·9	3·9676	·09387	·09057	·08741	·08437	·08145	·07865	·07595	·07335	·07085	·06844
4·0	3·4770	·11447	·11059	·10687	·10328	·09983	·09650	·09329	·09020	·08721	·08432
4·1	3·0665	·13701	·13255	·12825	·12410	·12010	·11623	·11249	·10888	·10538	·10200
4·2	2·7206	·16106	·15603	·15116	·14646	·14191	·13751	·13324	·12910	·12509	·12120
4·3	2·4276	·18608	·18051	·17512	·16989	·16481	·15989	·15511	·15046	·14595	·14156
4·4	2·1780	·21146	·20541	·19953	·19382	·18826	·18285	·17758	·17246	·16747	·16261
4·5	1·9640	·23655	·23008	·22377	·21763	·21164	·20580	·20010	·19454	·18911	·18382
4·6	1·7797	·26066	·25384	·24719	·24069	·23433	·22812	·22205	·21611	·21031	·20462
4·7	1·6202	·28315	·27607	·26914	·26236	·25572	·24921	·24283	·23659	·23046	·22446
4·8	1·4814	·30342	·29617	·28905	·28206	·27521	·26848	·26188	·25539	·24902	·24276
4·9	1·3599	·32096	·31361	·30638	·29928	·29229	·28542	·27866	·27201	·26547	·25904
5·0	1·2533	·33532	·32795	·32070	·31356	·30652	·29958	·29275	·28602	·27938	·27284
5·1	1·1593	·34619	·33889	·33168	·32457	·31756	·31063	·30380	·29705	·29039	·28381
5·2	1·0759	·35336	·34618	·33909	·33209	·32516	·31832	·31155	·30486	·29824	·29170
5·3	1·0018	·35672	·34974	·34282	·33598	·32921	·32250	·31587	·30930	·30279	·29635
5·4	0·9357	·35629		·34286	·33624	·32967	·32317	·31672	·31032	·30399	·29770
5·5	0·8764	·35217	·34571	·33930	·33294	·32663	·32037	·31416	·30799	·30187	·29580
5·6	0·8230	·34457	·33843	·33233	·32628	·32026	·31428	·30835	·30245	·29659	·29077
5·7	0·7749	·33377	·32798	·32222	·31650	·31081	·30515	·29953	·29393	·28837	·28284
5·8	0·7313	·32012	·31470	·30931	·30394	·29860	·29329	·28800	·28273	·27750	·27229
5·9	0·6917	·30403	·29899	·29398	·28898	·28401	·27906	·27413	·26921	·26432	·25945
6·0	0·6557	·28594	·28129	·27666	·27205	·26746	·26288	·25831	·25377	·24923	·24472

6·1	0·6227	·26631	·26206	·25782	·25359	·24938	·24518	·24099	·23681	·23265	·22850
6·2	0·5926	·24561	·24175	·23790	·23406	·23022	·22640	·22259	·21878	·21499	·21120
6·3	0·5649	·22431	·22083	·21736	·21390	·21044	·20699	·20354	·20010	·19667	·19325
6·4	0·5394	·20285	·19974	·19663	·19353	·19044	·18735	·18426	·18119	·17811	·17504
6·5	0·5158	·18163	·17887	·17612	·17337	·17062	·16788	·16514	·16240	·15967	·15695
6·6	0·4940	·16101	·15858	·15616	·15374	·15133	·14891	·14650	·14410	·14169	·13929
6·7	0·4739	·14129	·13918	·13707	·13496	·13286	·13075	·12865	·12655	·12445	·12236
6·8	0·4551	·12274	·12092	·11910	·11728	·11546	·11364	·11182	·11001	·10819	·10638
6·9	0·4376	·10554	·10398	·10242	·10086	·09931	·09775	·09619	·09464	·09309	·09153
7·0	0·4214	·08982	·08850	·08718	·08586	·08454	·08322	·08190	·08058	·07926	·07794
7·1	0·4062	·07565	·07454	·07343	·07232	·07122	·07011	·06900	·06789	·06679	·06568
7·2	0·3919	·06306	·06214	·06121	·06029	·05937	·05845	·05753	·05661	·05569	·05477
7·3	0·3786	·05201	·05125	·05049	·04974	·04898	·04822	·04746	·04670	·04594	·04519
7·4	0·3661	·04245	·04183	·04122	·04060	·03998	·03936	·03874	·03812	·03751	·03689
7·5	0·3543	·03429	·03379	·03329	·03279	·03229	·03179	·03129	·03079	·03030	·02980
7·6	0·3432	·02740	·02700	·02660	·02620	·02581	·02541	·02501	·02461	·02421	·02382
7·7	0·3327	·02166	·02135	·02103	·02072	·02041	·02009	·01978	·01946	·01915	·01883
7·8	0·3228	·01695	·01670	·01646	·01621	·01597	·01572	·01547	·01523	·01498	·01474
7·9	0·3134	·01312	·01293	·01274	·01255	·01236	·01217	·01198	·01179	·01160	·01141
8·0	0·3046	·01005	·00990	·00976	·00961	·00947	·00932	·00917	·00903	·00888	·00874
8·1	0·2962	·00762	·00750	·00739	·00728	·00717	·00706	·00695	·00684	·00673	·00662
8·2	0·2882	·00571	·00563	·00554	·00546	·00538	·00530	·00521	·00513	·00505	·00497
8·3	0·2806	·00424	·00418	·00411	·00405	·00399	·00393	·00387	·00381	·00375	·00368
8·4	0·2734	·00311	·00307	·00302	·00298	·00293	·00289	·00284	·00279	·00275	·00270
8·5	0·2666	·00226	·00223	·00219	·00216	·00213	·00210	·00206	·00203	·00200	·00196
8·6	0·2600	·00162	·00160	·00158	·00155	·00153	·00151	·00148	·00146	·00144	·00141
8·7	0·2538	·00115	·00114	·00112	·00110	·00109	·00107	·00105	·00104	·00102	·00100
8·8	0·2478	·00081	·00080	·00079	·00078	·00077	·00075	·00074	·00073	·00072	·00071
8·9	0·2421	·00057	·00056	·00055	·00054	·00053	·00053	·00052	·00051	·00050	·00049
9·0	0·2367	·00039	·00038	·00038	·00037	·00037	·00036	·00036	·00035	·00034	·00034

I am indebted to the Editors of the *Annals of Applied Biology* for permission to reproduce the first two sections of this table.

TABLE III. Maximum and Minimum Working Probits and Range

Minimum working probits			Maximum working probits	
Expected probit Y	$y_0 = Y - P/Z$	Range $1/Z$	$y_{100} = Y + Q/Z$	Expected probit Y
1·1	0·8579	5034	9·1421	8·9
1·2	0·9522	3425	9·0478	8·8
1·3	1·0462	2354	8·9538	8·7
1·4	1·1400	1634	8·8600	8·6
1·5	1·2334	1146	8·7666	8·5
1·6	1·3266	811·5	8·6734	8·4
1·7	1·4194	580·5	8·5806	8·3
1·8	1·5118	419·4	8·4882	8·2
1·9	1·6038	306·1	8·3962	8·1
2·0	1·6954	225·6	8·3046	8·0
2·1	1·7866	168·00	8·2134	7·9
2·2	1·8772	126·34	8·1228	7·8
2·3	1·9673	95·96	8·0327	7·7
2·4	2·0568	73·62	7·9432	7·6
2·5	2·1457	57·05	7·8543	7·5
2·6	2·2339	44·654	7·7661	7·4
2·7	2·3214	35·302	7·6786	7·3
2·8	2·4081	28·189	7·5919	7·2
2·9	2·4938	22·736	7·5062	7·1
3·0	2·5786	18·5216	7·4214	7·0
3·1	2·6624	15·2402	7·3376	6·9
3·2	2·7449	12·6662	7·2551	6·8
3·3	2·8261	10·6327	7·1739	6·7
3·4	2·9060	9·0154	7·0940	6·6
3·5	2·9842	7·7210	7·0158	6·5
3·6	3·0606	6·6788	6·9394	6·4
3·7	3·1351	5·8354	6·8649	6·3
3·8	3·2074	5·1497	6·7926	6·2
3·9	3·2773	4·5903	6·7227	6·1
4·0	3·3443	4·1327	6·6557	6·0
4·1	3·4083	3·7582	6·5917	5·9
4·2	3·4687	3·4519	6·5313	5·8
4·3	3·5251	3·2025	6·4749	5·7
4·4	3·5770	3·0010	6·4230	5·6
4·5	3·6236	2·8404	6·3764	5·5
4·6	3·6643	2·7154	6·3357	5·4
4·7	3·6982	2·6220	6·3018	5·3
4·8	3·7241	2·5573	6·2759	5·2
4·9	3·7407	2·5192	6·2593	5·1
5·0	3·7467	2·5066	6·2533	5·0
5·1	3·7401	2·5192	6·2599	4·9
5·2	3·7186	2·5573	6·2814	4·8
5·3	3·6798	2·6220	6·3202	4·7
5·4	3·6203	2·7154	6·3797	4·6
5·5	3·5360	2·8404	6·4640	4·5
5·6	3·4220	3·0010	6·5780	4·4
5·7	3·2724	3·2025	6·7276	4·3
5·8	3·0794	3·4519	6·9206	4·2
5·9	2·8335	3·7582	7·1665	4·1
6·0	2·5230	4·1327	7·4770	4·0
6·1	2·1324	4·5903	7·8676	3·9
6·2	1·6429	5·1497	8·3571	3·8
6·3	1·0295	5·8354	8·9705	3·7
6·4	0·2606	6·6788	9·7394	3·6
6·5	−0·7052	7·7210	10·7052	3·5

The working probit, y, may be obtained as $y = (Y - P/Z) + p/Z$ or $y = (Y + Q/Z) - q/Z$, whichever is the more convenient, where $p(= 1 - q)$ is the observed proportion killed.

TABLE IV. Working Probits

($Y = 2 \cdot 0$–$2 \cdot 9$; 0–50% kill)

% kill	Provisional probit, Y									
	2·0	2·1	2·2	2·3	2·4	2·5	2·6	2·7	2·8	2·9
0	1·695	1·787	1·877	1·967	2·057	2·146	2·234	2·321	2·408	2·494
1	3·951	3·467	3·141	2·927	2·793	2·716	2·681	2·674	2·690	2·721
2	6·207	5·147	4·404	3·886	3·529	3·287	3·127	3·027	·972	·949
3	8·463	6·827	5·667	4·846	4·265	·857	·574	·380	3·254	3·176
4	—	8·507	6·931	5·806	5·002	4·428	4·020	·733	·536	·403
5	—	—	8·194	6·765	·738	·998	·467	4·086	·818	·631
6	—	—	9·458	7·725	6·474	5·569	4·913	4·440	4·099	3·858
7	—	—	—	8·684	7·210	6·139	5·360	·793	·381	4·085
8	—	—	—	9·644	·946	·710	·806	5·146	·663	·313
9	—	—	—	—	8·683	7·280	6·253	·499	·945	·540
10	—	—	—	—	9·419	·851	·699	·852	5·227	·767
11	—	—	—	—	—	8·421	7·146	6·205	5·509	4·995
12	—	—	—	—	—	·992	·592	·558	·791	5·222
13	—	—	—	—	—	9·562	8·039	·911	6·073	·449
14	—	—	—	—	—	—	·486	7·264	·355	·677
15	—	—	—	—	—	—	·932	·617	·636	·904
16	—	—	—	—	—	—	9·379	7·970	6·918	6·132
17	—	—	—	—	—	—	·825	8·323	7·200	·359
18	—	—	—	—	—	—	—	·676	·482	·586
19	—	—	—	—	—	—	—	9·029	·764	·814
20	—	—	—	—	—	—	—	·382	8·046	7·041
21	—	—	—	—	—	—	—	9·735	8·328	7·268
22	—	—	—	—	—	—	—	—	·610	·496
23	—	—	—	—	—	—	—	—	·892	·723
24	—	—	—	—	—	—	—	—	9·173	·950
25	—	—	—	—	—	—	—	—	·455	8·178
26	—	—	—	—	—	—	—	—	9·737	8·405
27	—	—	—	—	—	—	—	—	—	·633
28	—	—	—	—	—	—	—	—	—	·860
29	—	—	—	—	—	—	—	—	—	9·087
30	—	—	—	—	—	—	—	—	—	·315
31	—	—	—	—	—	—	—	—	—	9·542
32	—	—	—	—	—	—	—	—	—	·769
33	—	—	—	—	—	—	—	—	—	·997
34	—	—	—	—	—	—	—	—	—	—
35	—	—	—	—	—	—	—	—	—	—

TABLE IV (cont.)

($Y = 3 \cdot 0$–$3 \cdot 9$; 0–50% kill)

% kill	Provisional probit, Y									
	3·0	3·1	3·2	3·3	3·4	3·5	3·6	3·7	3·8	3·9
0	2·579	2·662	2·745	2·826	2·906	2·984	3·061	3·135	3·207	3·277
1	2·764	2·815	2·872	2·932	2·996	3·061	3·127	3·193	3·259	3·323
2	·949	·967	·998	3·039	3·086	·139	·194	·252	·310	·369
3	3·134	3·120	3·125	·145	·176	·216	·261	·310	·362	·415
4	·319	·272	·252	·251	·267	·293	·328	·369	·413	·461
5	·505	·424	·378	·358	·357	·370	·395	·427	·465	·507
6	3·690	3·577	3·505	3·464	3·447	3·447	3·461	3·485	3·516	3·553
7	·875	·729	·632	·570	·537	·525	·528	·544	·568	·599
8	4·060	·882	·758	·677	·627	·602	·595	·602	·619	·645
9	·246	4·034	·885	·783	·717	·679	·662	·660	·671	·690
10	·431	·186	4·012	·889	·808	·756	·728	·719	·722	·736
11	4·616	4·339	4·138	3·996	3·898	3·834	3·795	3·777	3·774	3·782
12	·801	·491	·265	4·102	·988	·911	·862	·835	·825	·828
13	·986	·644	·391	·208	4·078	·988	·929	·894	·877	·874
14	5·172	·796	·518	·315	·168	4·065	·996	·952	·928	·920
15	·357	·948	·645	·421	·258	·142	4·062	4·010	·980	·966
16	5·542	5·101	4·771	4·527	4·348	4·220	4·129	4·069	4·031	4·012
17	·727	·253	·898	·634	·439	·297	·196	·127	·083	·058
18	·913	·406	5·025	·740	·529	·374	·263	·185	·134	·104
19	6·098	·558	·151	·846	·619	·451	·330	·244	·186	·149
20	·283	·710	·278	·953	·709	·528	·396	·302	·237	·195
21	6·468	5·863	5·405	5·059	4·799	4·606	4·463	4·361	4·289	4·241
22	·653	6·015	·531	·165	·889	·683	·530	·419	·340	·287
23	·839	·168	·658	·272	·979	·760	·597	·477	·392	·333
24	7·024	·320	·785	·378	5·070	·837	·664	·536	·443	·379
25	·209	·472	·911	·484	160	·914	·730	·594	·495	·425
26	7·394	6·625	6·038	5·591	5·250	4·992	4·797	4·652	4·546	4·471
27	·580	·777	·165	·697	·340	5·069	·864	·711	·598	·517
28	·765	·930	·291	·803	·430	·146	·931	·769	·649	·563
29	·950	7·082	·418	·910	·520	·223	·997	·827	·701	·608
30	8·135	·234	·545	6·016	·610	·300	5·064	·886	·752	·654
31	8·320	7·387	6·671	6·122	5·701	5·378	5·131	4·944	4·804	4·700
32	·506	·539	·798	·229	·791	·455	·198	5·002	·855	·746
33	·691	·692	·925	·335	·881	·532	·265	·061	·907	·792
34	·876	·844	7·051	·441	·971	·609	·331	·119	·958	·838
35	9·061	·996	·178	·548	6·061	·687	·398	·177	5·010	·884
36	9·247	8·149	7·305	6·654	6·151	5·764	5·465	5·236	5·061	4·930
37	·432	·301	·431	·760	·242	·841	·532	·294	·113	·976
38	·617	·454	·558	·867	·332	·918	·599	·353	·164	5·022
39	·802	·606	·685	·973	·422	·995	·665	·411	·216	·068
40	·987	·758	·811	7·079	·512	6·073	·732	·469	·267	·113
41	—	8·911	7·938	7·186	6·602	6·150	5·799	5·528	5·319	5·159
42	—	9·063	8·065	·292	·692	·227	·866	·586	·370	·205
43	—	·216	·191	·398	·782	·304	·932	·644	·422	·251
44	—	·368	·318	·505	·873	·381	·999	·703	·473	·297
45	—	·520	·445	·611	·963	·459	6·066	·761	·525	·343
46	—	9·673	8·571	7·717	7·053	6·536	6·133	5·819	5·576	5·389
47	—	·825	·698	·824	·143	·613	·200	·878	·628	·435
48	—	·978	·825	·930	·233	·690	·266	·936	·679	·481
49	—	—	·951	8·036	·323	·767	·333	·994	·731	·527
50	—	—	9·078	·143	·414	·845	·400	6·053	·782	·572

TABLE IV (*cont.*)

($Y = 4\cdot0\text{–}4\cdot9$; 0–50 % kill)

% kill	4·0	4·1	4·2	4·3	4·4	4·5	4·6	4·7	4·8	4·9
0	3·344	3·408	3·469	3·525	3·577	3·624	3·664	3·698	3·724	3·741
1	3·386	3·446	3·503	3·557	3·607	3·652	3·691	3·724	3·750	3·766
2	·427	·487	·538	·589	·637	·680	·719	·751	·775	·791
3	·468	·521	·572	·621	·667	·709	·746	·777	·801	·816
4	·510	·559	·607	·653	·697	·737	·773	·803	·826	·841
5	·551	·596	·641	·685	·727	·766	·800	·829	·852	·867
6	3·592	3·634	3·676	3·717	3·757	3·794	3·827	3·856	3·878	3·892
7	·634	·671	·710	·749	·787	·822	·854	·882	·903	·917
8	·675	·709	·745	·781	·817	·851	·882	·908	·929	·942
9	·716	·747	·779	·813	·847	·879	·909	·934	·954	·967
10	·758	·784	·814	·845	·877	·908	·936	·960	·980	·993
11	3·799	3·822	3·848	3·877	3·907	3·936	3·963	3·987	4·005	4·018
12	·840	·859	·883	·909	·937	·964	·990	4·013	·031	·043
13	·882	·897	·917	·941	·967	·993	4·017	·039	·057	·068
14	·923	·934	·952	·973	·997	4·021	·044	·065	·082	·093
15	·964	·972	·986	4·005	4·027	·050	·072	·092	·108	·119
16	4·006	4·010	4·021	4·038	4·057	4·078	4·099	4·118	4·133	4·144
17	·047	·047	·056	·070	·087	·106	·126	·144	·159	·169
18	·088	·085	·090	·102	·117	·135	·153	·170	·184	·194
19	·130	·122	·125	·134	·147	·163	·180	·196	·210	·219
20	·171	·160	·159	·166	·177	·192	·207	·223	·236	·245
21	4·212	4·198	4·194	4·198	4·207	4·220	4·235	4·249	4·261	4·270
22	·253	·235	·228	·230	·237	·248	·262	·275	·287	·295
23	·295	·273	·263	·262	·267	·277	·289	·301	·312	·320
24	·336	·310	·297	·294	·297	·305	·316	·327	·338	·345
25	·377	·348	·332	·326	·327	·334	·343	·354	·363	·370
26	4·419	4·385	4·366	4·358	4·357	4·362	4·370	4·380	4·389	4·396
27	·460	·423	·401	·390	·387	·391	·397	·406	·415	·421
28	·501	·461	·435	·422	·417	·419	·425	·432	·440	·446
29	·543	·498	·470	·454	·447	·447	·452	·459	·466	·471
30	·584	·536	·504	·486	·477	·476	·479	·485	·491	·496
31	4·625	4·573	4·539	4·518	4·507	4·504	4·506	4·511	4·517	4·522
32	·667	·611	·573	·550	·537	·533	·533	·537	·542	·547
33	·708	·649	·608	·582	·567	·561	·560	·563	·568	·572
34	·749	·686	·642	·614	·597	·589	·588	·590	·594	·597
35	·791	·724	·677	·646	·627	·618	·615	·616	·619	·622
36	4·832	4·761	4·711	4·678	4·657	4·646	4·642	4·642	4·645	4·648
37	·873	·799	·746	·710	·687	·675	·669	·668	·670	·673
38	·915	·836	·780	·742	·717	·703	·696	·695	·696	·698
39	·956	·874	·815	·774	·747	·731	·723	·721	·721	·723
40	·997	·912	·849	·806	·777	·760	·750	·747	·747	·748
41	5·039	4·949	4·884	4·838	4·807	4·788	4·778	4·773	4·773	4·774
42	·080	·987	·918	·870	·837	·817	·805	·799	·798	·799
43	·121	5·024	·953	·902	·867	·845	·832	·826	·824	·824
44	·163	·062	·988	·934	·897	·873	·859	·852	·849	·849
45	·204	·899	5·022	·966	·927	·902	·886	·878	·875	·874
46	5·245	5·137	5·057	4·998	4·957	4·930	4·913	4·904	4·900	4·900
47	·287	·175	·091	5·030	·987	·959	·941	·931	·926	·925
48	·328	·212	·126	·062	5·017	·987	·968	·957	·952	·950
49	·369	·250	·160	·094	·047	5·015	·995	·983	·977	·975
50	·411	·287	·195	·126	·078	·044	5·022	5·009	5·003	5·000

TABLE IV (cont.)

($Y = 5·0–5·9$; $0–50\%$ kill)

% kill	Provisional probit, Y									
	5·0	5·1	5·2	5·3	5·4	5·5	5·6	5·7	5·8	5·9
0	3·747	3·740	3·719	3·680	3·620	3·536	3·422	3·272	3·079	2·834
1	3·772	3·765	3·744	3·706	3·647	3·564	3·452	3·304	3·114	2·871
2	·797	·790	·770	·732	·675	·593	·482	·336	·148	·909
3	·822	·816	·795	·758	·702	·621	·512	·368	·183	·946
4	·847	·841	·821	·785	·729	·650	·542	·400	·217	·984
5	·872	·866	·846	·811	·756	·678	·572	·433	·252	3·021
6	3·897	3·891	3·872	3·837	3·783	3·706	3·602	3·465	3·287	3·059
7	·922	·916	·898	·863	·810	·735	·632	·497	·321	·097
8	·947	·942	·923	·890	·838	·763	·662	·529	·356	·134
9	·972	·967	·949	·916	·865	·792	·692	·561	·390	·172
10	·997	·992	·974	·942	·892	·820	·722	·593	·425	·209
11	4·022	4·017	4·000	3·968	3·919	3·848	3·752	3·625	3·459	3·247
12	·047	·042	·025	·994	·946	·877	·782	·657	·494	·284
13	·073	·068	·051	4·021	·973	·905	·812	·689	·528	·322
14	·098	·093	·077	·047	4·000	·934	·842	·721	·563	·360
15	·123	·118	·102	·073	·028	·962	·872	·753	·597	·397
16	4·148	4·143	4·128	4·099	4·055	3·990	3·902	3·785	3·632	3·435
17	·173	·168	·153	·126	·082	4·019	·932	·817	·666	·472
18	·198	·194	·179	·152	·109	·047	·962	·849	·701	·510
19	·223	·219	·204	·178	·136	·076	·992	·881	·735	·548
20	·248	·244	·230	·204	·163	·104	4·022	·913	·770	·585
21	4·273	4·269	4·256	4·230	4·191	4·132	4·052	3·945	3·804	3·623
22	·298	·294	·281	·257	·218	·161	·082	·977	·839	·660
23	·323	·320	·307	·283	·245	·189	·112	4·009	·873	·698
24	·348	·345	·332	·309	·272	·218	·142	·041	·908	·735
25	·373	·370	·358	·335	·299	·246	·172	·073	·942	·773
26	4·398	4·395	4·383	4·362	4·326	4·275	4·202	4·105	3·977	3·811
27	·423	·420	·409	·388	·353	·303	·232	·137	4·011	·848
28	·449	·445	·435	·414	·381	·331	·262	·169	·046	·886
29	·474	·471	·460	·440	·408	·360	·292	·201	·080	·923
30	·499	·496	·486	·466	·435	·388	·322	·233	·115	·961
31	4·524	4·521	4·511	4·493	4·462	4·417	4·352	4·265	4·149	3·999
32	·549	·546	·537	·519	·489	·445	·382	·297	·184	4·036
33	·574	·571	·563	·545	·516	·473	·412	·329	·219	·074
34	·599	·597	·588	·571	·544	·502	·442	·361	·253	·111
35	·624	·622	·614	·598	·571	·530	·472	·393	·288	·149
36	4·649	4·647	4·639	4·624	4·598	4·559	4·502	4·425	4·322	4·186
37	·674	·672	·665	·650	·625	·587	·532	·457	·357	·224
38	·699	·697	·690	·676	·652	·615	·562	·489	·391	·262
39	·724	·723	·716	·702	·679	·644	·592	·521	·426	·299
40	·749	·748	·742	·729	·706	·672	·622	·553	·460	·337
41	4·774	4·773	4·767	4·755	4·734	4·701	4·652	4·585	4·495	4·374
42	·799	·798	·793	·781	·761	·729	·682	·617	·529	·412
43	·825	·823	·818	·807	·788	·757	·712	·649	·564	·450
44	·850	·849	·844	·833	·815	·786	·742	·682	·598	·487
45	·875	·874	·869	·860	·842	·814	·772	·714	·633	·525
46	4·900	4·899	4·895	4·886	4·869	4·843	4·802	4·746	4·667	4·562
47	·925	·924	·921	·912	·897	·871	·832	·778	·702	·600
48	·950	·949	·946	·938	·924	·899	·862	·810	·736	·637
49	·975	·975	·972	·965	·951	·928	·892	·842	·771	·675
50	5·000	5·000	·997	·991	·978	·956	·922	·874	·805	·713

TABLE IV (cont.)

(Y = 6·0–6·9; 0–50% kill)

% kill	6·0	6·1	6·2	6·3	6·4	6·5	6·6	6·7	6·8	6·9
0	2·523	2·132	1·643	1·030	0·261	—	—	—	—	—
1	2·564	2·178	1·694	1·088	0·327	—	—	—	—	—
2	·606	·224	·746	·146	·394	—	—	—	—	—
3	·647	·270	·797	·205	·461	—	—	—	—	—
4	·688	·316	·849	·263	·528	—	—	—	—	—
5	·730	·362	·900	·321	·595	—	—	—	—	—
6	2·771	2·408	1·952	1·380	0·661	—	—	—	—	—
7	·812	·454	2·003	·438	·728	—	—	—	—	—
8	·854	·500	·055	·496	·795	—	—	—	—	—
9	·895	·546	·106	·555	·862	—	—	—	—	—
10	·936	·591	·158	·613	·928	0·067	—	—	—	—
11	2·978	2·637	2·209	1·671	0·995	0·144	—	—	—	—
12	3·019	·683	·261	·730	1·062	·221	—	—	—	—
13	·060	·729	·312	·788	·129	·299	—	—	—	—
14	·102	·775	·364	·846	·196	·376	—	—	—	—
15	·143	·821	·415	·905	·262	·453	—	—	—	—
16	3·184	2·867	2·467	1·963	1·329	0·530	—	—	—	—
17	·226	·913	·518	2·022	·396	·607	—	—	—	—
18	·267	·959	·570	·080	·463	·685	—	—	—	—
19	·308	3·005	·621	·138	·530	·762	—	—	—	—
20	·350	·050	·673	·197	·596	·839	—	—	—	—
21	3·391	3·096	2·724	2·255	1·663	0·916	—	—	—	—
22	·432	·142	·776	·313	·730	·993	0·062	—	—	—
23	·474	·188	·827	·372	·797	1·071	·152	—	—	—
24	·515	·234	·879	·430	·864	·148	·243	—	—	—
25	·556	·280	·930	·488	·930	·225	·333	—	—	—
26	3·598	3·326	2·982	2·547	1·997	1·302	0·423	—	—	—
27	·639	·372	3·033	·605	2·064	·379	·513	—	—	—
28	·680	·418	·085	·663	·131	·457	·603	—	—	—
29	·721	·464	·136	·722	·197	·534	·693	—	—	—
30	·763	·509	·188	·780	·264	·611	·784	—	—	—
31	3·804	3·555	3·239	2·838	2·331	1·688	0·874	—	—	—
32	·845	·601	·291	·897	·398	·766	·964	—	—	—
33	·887	·647	·342	·955	·465	·843	1·054	0·050	—	—
34	·928	·693	·394	3·014	·531	·920	·144	·156	—	—
35	·969	·739	·445	·072	·598	·997	·234	·262	—	—
36	4·011	3·785	3·497	3·130	2·665	2·074	1·324	0·369	—	—
37	·052	·831	·548	·189	·732	·152	·415	·475	—	—
38	·093	·877	·600	·247	·799	·229	·505	·581	—	—
39	·135	·923	·651	·305	·865	·306	·595	·688	—	—
40	·176	·969	·703	·364	·932	·383	·685	·794	—	—
41	4·217	4·014	3·754	3·422	2·999	2·460	1·775	0·900	—	—
42	·259	·060	·806	·480	3·066	·538	·865	1·007	—	—
43	·300	·106	·857	·539	·132	·615	·955	·113	0·035	—
44	·341	·152	·909	·597	·199	·692	2·046	·219	·162	—
45	·383	·198	·960	·655	·266	·769	·136	·326	·289	—
46	4·424	4·244	4·012	3·714	3·333	2·846	2·226	1·432	0·415	—
47	·465	·290	·063	·772	·400	·924	·316	·538	·542	—
48	·507	·336	·115	·830	·466	3·001	·406	·645	·669	—
49	·548	·382	·166	·889	·533	·078	·496	·751	·795	—
50	·589	·428	·218	·947	·600	·155	·586	·857	·922	—

TABLE IV (cont.)

($Y = 3\cdot0-3\cdot9$; 51–100 % kill)

% kill	3·0	3·1	3·2	3·3	3·4	3·5	3·6	3·7	3·8	3·9
					Provisional probit, Y					
51	—	—	9·205	8·249	7·504	6·922	6·467	6·111	5·834	5·618
52	—	—	·331	·355	·594	·999	·534	·170	·885	·664
53	—	—	·458	·462	·684	7·076	·600	·228	·937	·710
54	—	—	·585	·568	·774	·154	·667	·286	·988	·756
55	—	—	·711	·674	·864	·231	·734	·345	6·040	·802
56	—	—	9·838	8·781	7·954	7·308	6·801	6·403	6·091	5·848
57	—	—	·965	·887	8·045	·385	·868	·461	·143	·894
58	—	—	—	·993	·135	·462	·934	·520	·194	·940
59	—	—	—	9·100	·225	·540	7·001	·578	·246	·986
60	—	—	—	·206	·315	·617	·068	·636	·297	6·031
61	—	—	—	9·312	8·405	7·694	7·135	6·695	6·349	6·077
62	—	—	—	·419	·495	·771	·201	·753	·400	·123
63	—	—	—	·525	·585	·848	·268	·811	·452	·169
64	—	—	—	·631	·676	·926	·335	·870	·503	·215
65	—	—	—	·738	·766	8·003	·402	·928	·555	·261
66	—	—	—	9·844	8·856	8·080	7·469	6·986	6·606	6·307
67	—	—	—	·950	·946	·157	·535	7·045	·658	·353
68	—	—	—	—	9·036	·234	·602	·103	·709	·399
69	—	—	—	—	·126	·312	·669	·162	·761	·445
70	—	—	—	—	·216	·389	·736	·220	·812	·491
71	—	—	—	—	9·307	8·466	7·803	7·278	6·864	6·536
72	—	—	—	—	·397	·543	·869	·337	·915	·582
73	—	—	—	—	·487	·621	·936	·395	·967	·628
74	—	—	—	—	·577	·698	8·003	·453	7·018	·674
75	—	—	—	—	·667	·775	·070	·512	·070	·720
76	—	—	—	—	9·757	8·852	8·136	7·570	7·121	6·766
77	—	—	—	—	·848	·929	·203	·628	·173	·812
78	—	—	—	—	·938	9·007	·270	·687	·224	·858
79	—	—	—	—	—	·084	·337	·745	·276	·904
80	—	—	—	—	—	·161	·404	·803	·327	·950
81	—	—	—	—	—	9·238	8·470	7·862	7·379	6·995
82	—	—	—	—	—	·315	·537	·920	·430	7·041
83	—	—	—	—	—	·393	·604	·978	·482	·087
84	—	—	—	—	—	·470	·671	8·037	·533	·133
85	—	—	—	—	—	·547	·738	·095	·585	·179
86	—	—	—	—	—	9·624	8·804	8·154	7·636	7·225
87	—	—	—	—	—	·701	·871	·212	·688	·271
88	—	—	—	—	—	·779	·938	·270	·739	·317
89	—	—	—	—	—	·856	9·005	·329	·791	·363
90	—	—	—	—	—	·933	·072	·387	·842	·409
91	—	—	—	—	—	—	9·138	8·445	7·894	7·454
92	—	—	—	—	—	—	·205	·504	·945	·500
93	—	—	—	—	—	—	·272	·562	·997	·546
94	—	—	—	—	—	—	·339	·620	8·048	·592
95	—	—	—	—	—	—	·405	·679	·100	·638
96	—	—	—	—	—	—	9·472	8·737	8·151	7·684
97	—	—	—	—	—	—	·539	·795	·203	·730
98	—	—	—	—	—	—	·606	·854	·254	·776
99	—	—	—	—	—	—	·673	·912	·306	·822
100	—	—	—	—	—	—	·739	·970	·357	·868

TABLE IV (*cont.*)

($Y = 4·0–4·9$; 51–100 % kill)

% kill	4·0	4·1	4·2	4·3	4·4	4·5	4·6	4·7	4·8	4·9
				Provisional probit, Y						
51	5·452	5·325	5·229	5·158	5·108	5·072	5·049	5·035	5·028	5·025
52	·493	·363	·264	·190	·138	·101	·076	·062	·054	·051
53	·535	·400	·298	·222	·168	·129	·103	·088	·079	·076
54	·576	·438	·333	·254	·198	·157	·131	·114	·105	·101
55	·617	·475	·367	·286	·228	·186	·158	·140	·131	·126
56	5·659	5·513	5·402	5·318	5·258	5·214	5·185	5·167	5·156	5·151
57	·700	·550	·436	·351	·288	·243	·212	·193	·182	·177
58	·741	·588	·471	·383	·318	·271	·239	·219	·207	·202
59	·783	·626	·505	·415	·348	·299	·266	·245	·233	·227
60	·824	·663	·540	·447	·378	·328	·294	·271	·258	·252
61	5·865	5·701	5·574	5·479	5·408	5·356	5·321	5·298	5·284	5·277
62	·907	·738	·609	·511	·438	·385	·348	·324	·310	·303
63	·948	·776	·643	·543	·468	·413	·375	·350	·335	·328
64	·989	·814	·678	·575	·498	·441	·402	·376	·361	·353
65	6·031	·851	·712	·607	·528	·470·	·429	·402	·386	·378
66	6·072	5·889	5·747	5·639	5·558	5·498	5·456	5·429	5·412	5·403
67	·113	·926	·781	·671	·588	·527	·484	·455	·437	·429
68	·155	·964	·816	·703	·618	·555	·511	·481	·463	·454
69	·196	6·001	·851	·735	·648	·583	·538	·507	·489	·479
70	·237	·039	·885	·767	·678	·612	·565	·534	·514	·504
71	6·279	6·077	5·920	5·799	5·708	5·640	5·592	5·560	5·540	5·529
72	·320	·114	·954	·831	·738	·669	·619	·586	·565	·555
73	·361	·152	·989	·863	·768	·697	·647	·612	·591	·580
74	·402	·189	6·023	·895	·798	·725	·674	·638	·617	·605
75	·444	·227	·058	·927	·828	·754	·701	·665	·642	·630
76	6·485	6·265	6·092	5·959	5·858	5·782	5·728	5·691	5·668	5·655
77	·526	·302	·127	·991	·888	·811	·755	·717	·693	·680
78	·568	·340	·161	6·023	·918	·839	·782	·743	·719	·706
79	·609	·377	·196	·055	·948	·868	·809	·770	·744	·731
80	·650	·415	·230	·087	·978	·896	·837	·796	·770	·756
81	6·692	6·452	6·265	6·119	6·008	5·924	5·864	5·822	5·796	5·781
82	·733	·490	·299	·151	·038	·953	·891	·848	·821	·806
83	·774	·528	·334	·183	·068	·981	·918	·874	·847	·832
84	·816	·565	·368	·215	·098	6·010	·945	·901	·872	·857
85	·857	·603	·403	·247	·128	·038	·972	·927	·898	·882
86	6·898	6·640	6·437	6·279	6·158	6·066	6·000	5·953	5·923	5·907
87	·940	·678	·472	·311	·188	·095	·027	·979	·949	·932
88	·981	·716	·506	·343	·218	·123	·054	6·006	·975	·958
89	7·022	·753	·541	·375	·248	·152	·081	·032	6·000	·983
90	·064	·791	·575	·407	·278	·180	·108	·058	·026	6·008
91	7·105	6·828	6·610	6·439	6·308	6·208	6·135	6·084	6·051	6·033
92	·146	·866	·644	·471	·338	·237	·162	·110	·077	·058
93	·188	·903	·679	·503	·368	·265	·190	·137	·102	·084
94	·229	·941	·713	·535	·398	·294	·217	·163	·128	·109
95	·270	·979	·748	·567	·428	·322	·244	·189	·154	·134
96	7·312	7·016	6·783	6·600	6·458	6·350	6·271	6·215	6·179	6·159
97	·353	·054	·817	·632	·488	·379	·298	·242	·205	·184
98	·394	·091	·852	·664	·518	·407	·325	·268	·230	·210
99	·436	·129	·886	·696	·548	·436	·353	·294	·256	·235
100	·477	·166	·921	·728	·578	·464	·380	·320	·281	·260

TABLE IV (cont.)

$(Y = 5 \cdot 0 – 5 \cdot 9;\ 51–100\%\ \text{kill})$

% kill	\multicolumn{10}{c}{Provisional probit, Y}
	5·0	5·1	5·2	5·3	5·4	5·5	5·6	5·7	5·8	5·9
51	5·025	5·025	5·023	5·017	5·005	4·985	4·953	4·906	4·840	4·750
52	·050	·050	·048	·043	·032	5·013	·983	·938	·874	·788
53	·075	·075	·074	·069	·059	·041	5·013	·970	·909	·825
54	·100	·100	·100	·096	·087	·070	·043	5·002	·943	·863
55	·125	·126	·125	·122	·114	·098	·073	·034	·978	·901
56	5·150	5·151	5·151	5·148	5·141	5·127	5·103	5·066	5·012	4·938
57	·175	·176	·176	·174	·168	·155	·133	·098	·047	·976
58	·201	·201	·202	·201	·195	·183	·163	·130	·082	5·013
59	·226	·226	·227	·227	·222	·212	·193	·162	·116	·051
60	·251	·252	·253	·253	·250	·240	·223	·194	·151	·088
61	5·276	5·277	5·279	5·279	5·277	5·269	5·253	5·226	5·185	5·126
62	·301	·302	·304	·305	·304	·297	·283	·258	·220	·164
63	·326	·327	·330	·332	·331	·325	·313	·290	·254	·201
64	·351	·352	·355	·358	·358	·354	·343	·322	·289	·239
65	·376	·378	·381	·384	·385	·382	·373	·354	·323	·276
66	5·401	5·403	5·406	5·410	5·412	5·411	5·403	5·386	5·358	5·314
67	·426	·428	·432	·437	·440	·439	·433	·418	·392	·351
68	·451	·453	·458	·463	·467	·467	·463	·450	·427	·389
69	·476	·478	·483	·489	·494	·496	·493	·482	·461	·427
70	·501	·504	·509	·515	·521	·524	·523	·514	·496	·464
71	5·526	5·529	5·534	5·541	5·548	5·553	5·553	5·546	5·530	5·502
72	·551	·554	·560	·568	·575	·581	·583	·578	·565	·539
73	·577	·579	·585	·594	·603	·609	·613	·610	·599	·577
74	·602	·604	·611	·620	·630	·638	·643	·642	·634	·615
75	·627	·630	·637	·646	·657	·666	·673	·674	·668	·652
76	5·652	5·655	5·662	5·673	5·684	5·695	5·703	5·706	5·703	5·690
77	·677	·680	·688	·699	·711	·723	·733	·738	·737	·727
78	·702	·705	·713	·725	·738	·752	·763	·770	·772	·765
79	·727	·730	·739	·751	·765	·780	·793	·802	·806	·802
80	·752	·755	·764	·777	·793	·808	·823	·834	·841	·840
81	5·777	5·781	5·790	5·804	5·820	5·837	5·853	5·866	5·875	5·878
82	·802	·806	·816	·830	·847	·865	·883	·898	·910	·915
83	·827	·831	·841	·856	·874	·894	·913	·930	·944	·953
84	·852	·856	·867	·882	·901	·922	·943	·962	·979	·990
85	·877	·881	·892	·908	·928	·950	·973	·995	6·014	6·028
86	5·902	5·907	5·918	5·935	5·956	5·979	6·003	6·027	6·048	6·066
87	·927	·932	·943	·961	·983	6·007	·033	·059	·083	·103
88	·953	·957	·969	·987	6·010	·036	·063	·091	·117	·141
89	·978	·982	·995	6·013	·037	·064	·093	·123	·152	·178
90	6·003	6·007	6·020	·040	·064	·092	·123	·155	·186	·216
91	6·028	6·033	6·046	6·066	6·091	6·121	6·153	6·187	6·221	6·253
92	·053	·058	·071	·092	·118	·149	·183	·219	·255	·291
93	·078	·083	·097	·118	·146	·178	·213	·251	·290	·329
94	·103	·108	·122	·144	·173	·206	·243	·283	·324	·366
95	·128	·133	·148	·171	·200	·234	·273	·315	·359	·404
96	6·153	6·159	6·174	6·197	6·227	6·263	6·303	6·347	6·393	6·441
97	·178	·184	·199	·223	·254	·291	·333	·379	·428	·479
98	·203	·209	·225	·249	·281	·320	·363	·411	·462	·517
99	·228	·234	·250	·276	·309	·348	·393	·443	·497	·554
100	·253	·259	·276	·302	·336	·376	·423	·475	·531	·592

TABLE IV (cont.)

(Y = 6·0–6·9; 51–100% kill)

% kill	Provisional probit, Y									
	6·0	6·1	6·2	6·3	6·4	6·5	6·6	6·7	6·8	6·9
51	4·631	4·473	4·269	4·006	3·667	3·233	2·677	1·964	1·049	—
52	·672	·519	·321	·064	·734	·310	·767	2·070	·175	0·022
53	·713	·565	·372	·122	·800	·387	·857	·176	·302	·175
54	·755	·611	·424	·181	·867	·464	·947	·283	·429	·327
55	·796	·657	·475	·239	·934	·541	3·037	·389	·555	·480
56	4·837	4·703	4·527	4·297	4·001	3·619	3·127	2·495	1·682	0·632
57	·879	·749	·578	·356	·068	·696	·218	·602	·809	·784
58	·920	·795	·630	·414	·134	·773	·308	·708	·935	·937
59	·961	·841	·681	·472	·201	·850	·398	·814	2·062	1·089
60	5·003	·887	·733	·531	·268	·927	·488	·921	·189	·242
61	5·044	4·932	4·784	4·589	4·335	4·005	3·578	3·027	2·315	1·394
62	·085	·978	·836	·647	·401	·082	·668	·133	·442	·546
63	·127	5·024	·887	·706	·468	·159	·758	·240	·569	·699
64	·168	·070	·939	·764	·535	·236	·849	·346	·695	·851
65	·209	·116	·990	·823	·602	·313	·939	·452	·822	2·004
66	5·251	5·162	5·042	4·881	4·669	4·391	4·029	3·559	2·949	2·156
67	·292	·208	·093	·939	·735	·468	·119	·665	3·075	·308
68	·333	·254	·145	·998	·802	·545	·209	·771	·202	·461
69	·375	·300	·196	5·056	·869	·622	·299	·878	·329	·613
70	·416	·346	·248	·114	·936	·700	·390	·984	·455	·766
71	5·457	5·392	5·299	5·173	5·003	4·777	4·480	4·090	3·582	2·918
72	·499	·437	·351	·231	·069	·854	·570	·197	·709	3·070
73	·540	·483	·402	·289	·136	·931	·660	·303	·835	·223
74	·581	·529	·454	·348	·203	5·008	·750	·409	·962	·375
75	·623	·575	·505	·406	·270	·086	·840	·516	4·089	·528
76	5·664	5·621	5·557	5·464	5·336	5·163	4·930	4·622	4·215	3·680
77	·705	·667	·608	·523	·403	·240	5·021	·728	·342	·832
78	·747	·713	·660	·581	·470	·317	·111	·835	·469	·985
79	·788	·759	·711	·639	·537	·394	·201	·941	·595	4·137
80	·829	·805	·763	·698	·604	·472	·291	5·047	·722	·290
81	5·870	5·851	5·814	5·756	5·670	5·549	5·381	5·154	4·849	4·442
82	·912	·896	·866	·815	·737	·626	·471	·260	·975	·594
83	·953	·942	·917	·873	·804	·703	·561	·366	5·102	·747
84	·994	·988	·969	·931	·871	·780	·652	·473	·229	·899
85	6·036	6·034	6·020	·990	·938	·858	·742	·579	·355	5·052
86	6·077	6·080	6·072	6·048	6·004	5·935	5·832	5·685	5·482	5·204
87	·118	·126	·123	·106	·071	6·012	·922	·792	·609	·356
88	·160	·172	·175	·165	·138	·089	6·012	·898	·735	·509
89	·201	·218	·226	·223	·205	·166	·102	6·004	·862	·661
90	·242	·264	·278	·281	·272	·244	·192	·111	·988	·814
91	6·284	6·310	6·329	6·340	6·338	6·321	6·283	6·217	6·115	5·966
92	·325	·355	·381	·398	·405	·398	·373	·323	·242	6·118
93	·366	·401	·432	·456	·472	·475	·463	·430	·368	·271
94	·408	·447	·484	·515	·539	·553	·553	·536	·495	·423
95	·449	·493	·535	·573	·605	·630	·643	·642	·622	·576
96	6·490	6·539	6·587	6·631	6·672	6·707	6·733	6·749	6·748	6·728
97	·532	·585	·638	·690	·739	·784	·824	·855	·875	·880
98	·573	·631	·690	·748	·806	·861	·914	·961	7·002	7·033
99	·614	·677	·741	·807	·873	·939	7·004	7·068	·128	·185
100	·656	·723	·793	·865	·939	7·016	·094	·174	·255	·338

TABLE IV (cont.)

($Y = 7\cdot0$–$7\cdot9$; 51–100% kill)

% kill	7·0	7·1	7·2	7·3	7·4	7·5	7·6	7·7	7·8	7·9
					Provisional probit, Y					
51	—									
52	—	—	—	—	—	—	—	—	—	—
53	—	—	—	—	—	—	—	—	—	—
54	—	—	—	—	—	—	—	—	—	—
55	—	—	—	—	—	—	—	—	—	—
56	—	—	—	—	—	—	—	—	—	—
57	—	—	—	—	—	—	—	—	—	—
58	—	—	—	—	—	—	—	—	—	—
59	—	—	—	—	—	—	—	—	—	—
60	0·013	—	—	—	—	—	—	—	—	—
61	0·198	—	—	—	—	—	—	—	—	—
62	·383	—	—	—	—	—	—	—	—	—
63	·568	—	—	—	—	—	—	—	—	—
64	·753	—	—	—	—	—	—	—	—	—
65	·939	—	—	—	—	—	—	—	—	—
66	1·124	—	—	—	—	—	—	—	—	—
67	·309	0·003	—	—	—	—	—	—	—	—
68	·494	·231	—	—	—	—	—	—	—	—
69	·680	·458	—	—	—	—	—	—	—	—
70	·865	·685	—	—	—	—	—	—	—	—
71	2·050	0·913	—	—	—	—	—	—	—	—
72	·235	1·140	—	—	—	—	—	—	—	—
73	·420	·367	—	—	—	—	—	—	—	—
74	·606	·595	0·263	—	—	—	—	—	—	—
75	·791	·822	·545	—	—	—	—	—	—	—
76	2·976	2·050	0·827	—	—	—	—	—	—	—
77	3·161	·277	1·108	—	—	—	—	—	—	—
78	·347	·504	·390	—	—	—	—	—	—	—
79	·532	·732	·672	0·265	—	—	—	—	—	—
80	·717	·959	·954	·618	—	—	—	—	—	—
81	3·902	3·186	2·236	0·971	—	—	—	—	—	—
82	4·087	·414	·518	1·324	—	—	—	—	—	—
83	·273	·641	·800	·677	0·175	—	—	—	—	—
84	·458	·868	3·082	2·030	·621	—	—	—	—	—
85	·643	4·096	·364	·383	1·068	—	—	—	—	—
86	4·828	4·323	3·645	2·736	1·514	—	—	—	—	—
87	5·014	·551	·927	3·089	·961	0·438	—	—	—	—
88	·199	·778	4·209	·442	2·408	1·008	—	—	—	—
89	·384	5·005	·491	·795	·854	·579	—	—	—	—
90	·569	·233	·773	4·148	3·301	2·149	0·581	—	—	—
91	5·754	5·460	5·055	4·501	3·747	2·720	1·317	—	—	—
92	·940	·687	·337	·854	4·194	3·290	2·054	0·356	—	—
93	6·125	·915	·619	5·207	·640	·861	·790	1·316	—	—
94	·310	6·142	·901	·560	5·087	4·431	3·526	2·275	0·542	—
95	·495	·369	6·182	·914	·533	5·002	4·262	3·235	1·806	—
96	6·681	6·597	6·464	6·267	5·980	5·572	4·998	4·194	3·069	1·493
97	·866	·824	·746	·620	6·426	6·143	5·735	5·154	4·333	3·173
98	7·051	7·051	7·028	·973	·873	·713	6·471	6·114	5·596	4·853
99	·236	·279	·310	7·326	7·319	7·284	7·207	7·073	6·859	6·533
100	·421	·506	·592	·679	·766	·854	·943	8·033	8·123	8·213

TABLE V.   The Ordinate, Z, and Z^2

Y		Z	Z^2
5·0		0·39894	0·15915
4·9,	5·1	0·39695	0·15757
4·8,	5·2	0·39104	0·15291
4·7,	5·3	0·38139	0·14546
4·6,	5·4	0·36827	0·13562
4·5,	5·5	0·35207	0·12395
4·4,	5·6	0·33322	0·11104
4·3,	5·7	0·31225	0·09750
4·2,	5·8	0·28969	0·08392
4·1,	5·9	0·26609	0·07080
4·0,	6·0	0·24197	0·05855
3·9,	6·1	0·21785	0·04746
3·8,	6·2	0·19419	0·03771
3·7,	6·3	0·17137	0·02937
3·6,	6·4	0·14973	0·02242
3·5,	6·5	0·12952	0·01677
3·4,	6·6	0·11092	0·01230
3·3,	6·7	0·09405	0·00885
3·2,	6·8	0·07895	0·00623
3·1,	6·9	0·06562	0·00431
3·0,	7·0	0·05399	0·00292
2·9,	7·1	0·04398	0·00193
2·8,	7·2	0·03547	0·00126
2·7,	7·3	0·02833	0·00080
2·6,	7·4	0·02239	0·00050
2·5,	7·5	0·01753	0·00031
2·4,	7·6	0·01358	0·00018
2·3,	7·7	0·01042	0·00011
2·2,	7·8	0·00792	0·00006
2·1,	7·9	0·00595	0·00004
2·0,	8·0	0·00443	0·00002
1·9,	8·1	0·00327	0·00001
1·8,	8·2	0·00238	0·00001
1·7,	8·3	0·00172	0·00000
1·6,	8·4	0·00123	0·00000
1·5,	8·5	0·00087	0·00000
1·4,	8·6	0·00061	0·00000
1·3,	8·7	0·00042	0·00000
1·2,	8·8	0·00029	0·00000
1·1,	8·9	0·00020	0·00000
1·0,	9·0	0·00013	0·00000

TABLE VI. Distribution of χ^2

Degrees of freedom	Probability								
	·90	·70	·50	·30	·10	·05	·02	·01	·001
1	·016	·15	·45	1·1	2·7	3·8	5·4	6·6	10·8
2	·21	·71	1·4	2·4	4·6	6·0	7·8	9·2	13·8
3	·58	1·4	2·4	3·7	6·3	7·8	9·8	11·3	16·3
4	1·1	2·2	3·4	4·9	7·8	9·5	11·7	13·3	18·5
5	1·6	3·0	4·4	6·1	9·2	11·1	13·4	15·1	20·5
6	2·2	3·8	5·3	7·2	10·6	12·6	15·0	16·8	22·5
7	2·8	4·7	6·3	8·4	12·0	14·1	16·6	18·5	24·3
8	3·5	5·5	7·3	9·5	13·4	15·5	18·2	20·1	26·1
9	4·2	6·4	8·3	10·7	14·7	16·9	19·7	21·7	27·9
10	4·9	7·3	9·3	11·8	16·0	18·3	21·2	23·2	29·6
12	6·3	9·0	11·3	14·0	18·5	21·0	24·1	26·2	32·9
14	7·8	10·8	13·3	16·2	21·1	23·7	26·9	29·1	36·1
16	9·3	12·6	15·3	18·4	23·5	26·3	29·6	32·0	39·3
18	10·9	14·4	17·3	20·6	26·0	28·9	32·3	34·8	42·3
20	12·4	16·3	19·3	22·8	28·4	31·4	35·0	37·6	45·3
22	14·0	18·1	21·3	24·9	30·8	33·9	37·7	40·3	48·3
24	15·7	19·9	23·3	27·1	33·2	36·4	40·3	43·0	51·2
26	17·3	21·8	25·3	29·2	35·6	38·9	42·9	45·6	54·1
28	18·9	23·6	27·3	31·4	37·9	41·3	45·4	48·3	56·9
30	20·6	25·5	29·3	33·5	40·3	43·8	48·0	50·9	59·7

When χ^2 is based on more than 30 degrees of freedom, the quantity $\sqrt{(2\chi^2)} - \sqrt{(2f-1)}$ (where f is the number of degrees of freedom) has approximately the following distribution:

>30	−1·28	−0·52	0·00	0·52	1·28	1·64	2·05	2·33	3·09

I am indebted to Professor R. A. Fisher and Dr F. Yates, and also to Messrs Oliver and Boyd, Ltd., of Edinburgh, for permission to print Table VI as an abridgement of Table IV of their book *Statistical Tables for Biological, Agricultural and Medical Research*.

Table VII. Distribution of t

Degrees of freedom	Probability								
	·90	·70	·50	·30	·10	·05	·02	·01	·001
1	·16	·51	1·00	1·96	6·31	12·7	31·8	63·7	637·
2	·14	·44	·82	1·39	2·92	4·30	6·96	9·92	31·6
3	·14	·42	·76	1·25	2·35	3·18	4·54	5·84	12·9
4	·13	·41	·74	1·19	2·13	2·78	3·75	4·60	8·61
5	·13	·41	·73	1·16	2·02	2·57	3·36	4·03	6·86
6	·13	·40	·72	1·13	1·94	2·45	3·14	3·71	5·96
7	·13	·40	·71	1·12	1·90	2·36	3·00	3·50	5·40
8	·13	·40	·71	1·11	1·86	2·31	2·90	3·36	5·04
9	·13	·40	·70	1·10	1·83	2·26	2·82	3·25	4·78
10	·13	·40	·70	1·09	1·81	2·23	2·76	3·17	4·59
12	·13	·40	·70	1·08	1·78	2·18	2·68	3·06	4·32
14	·13	·39	·69	1·08	1·76	2·14	2·62	2·98	4·14
16	·13	·39	·69	1·07	1·75	2·12	2·58	2·92	4·02
18	·13	·39	·69	1·07	1·73	2·10	2·55	2·88	3·92
20	·13	·39	·69	1·06	1·72	2·09	2·53	2·84	3·85
22	·13	·39	·69	1·06	1·72	2·07	2·51	2·82	3·79
24	·13	·39	·68	1·06	1·71	2·06	2·49	2·80	3·74
26	·13	·39	·68	1·06	1·71	2·06	2·48	2·78	3·71
28	·13	·39	·68	1·06	1·70	2·05	2·47	2·76	3·67
30	·13	·39	·68	1·06	1·70	2·04	2·46	2·75	3·65
40	·13	·39	·68	1·05	1·68	2·02	2·42	2·70	3·55
60	·13	·39	·68	1·05	1·67	2·00	2·39	2·66	3·46
120	·13	·39	·68	1·04	1·66	1·98	2·36	2·62	3·37
∞	·126	·385	·674	1·036	1·645	1·960	2·326	2·576	3·291

I am indebted to Professor R. A. Fisher and Dr F. Yates, and also to Messrs Oliver and Boyd, Ltd., of Edinburgh, for permission to print Table VII as an abridgement of Table III of their book *Statistical Tables for Biological, Agricultural and Medical Research*.

INDEX OF AUTHORS

INDEX OF SUBJECTS